Hannah Yates
Dry Sandford

Collins New Primary Maths

Assessment Pack 2

Author: Peter Clarke

William Collins' dream of knowledge for all began with the publication of his first book in 1819. A self-educated mill worker, he not only enriched millions of lives, but also founded a flourishing publishing house. Today, staying true to this spirit, Collins books are packed with inspiration, innovation and practical expertise. They place you at the centre of a world of possibility and give you exactly what you need to explore it.

Collins. Freedom to Teach.

Published by Collins
An imprint of HarperCollinsPublishers
77 – 85 Fulham Palace Road
Hammersmith
London
W6 8JB

Browse the complete Collins catalogue at
www.collinseducation.com

10 9 8 7 6 5

ISBN-13 978 0 00 722032 8

Peter Clarke asserts his moral right to be identified as the author of this work

British Library Cataloguing in Publication Data
A Catalogue record for this publication is available from the British Library

Cover design by Laing&Carroll
Cover artwork by Jonatronix Ltd
Internal design and page make-up by Neil Adams
Illustrations by Neil Adams and Bridget Dowty
Edited by Ros and Chris Davies

Acknowledgement
The author would like to thank Brian Molyneaux for his valuable contribution to this product.

Printed and bound by Martins the Printers, Berwick-upon-Tweed

Contents

Introduction

What does the Primary National Strategy (PNS) *Renewed Framework for Mathematics* (2006) say about assessment?

The PNS *Renewed Framework for Mathematics* identifies two main purposes of assessment:

- Assessment *for* learning (formative on-going assessment)
- Assessment *of* learning (summative assessment)

Assessment *for* learning involves both pupils and teachers finding out about the specific strengths and weaknesses of individual children, and the class as a whole, and using this to inform future teaching and learning.

Assessment *for* learning:

– is part of the planning process
– is informed by learning objectives
– engages children in the assessment process
– recognises the achievements of all children
– takes account of how children learn
– motivates learners.

Assessment *of* learning is any assessment that summarises where individual children, and the class as a whole, are at a given point in time. It provides a snapshot of what has been learned.

The *Collins NEW Primary Maths (CNPM) Assessment Packs*

The *CNPM Assessment Packs* aim to provide guidance in both Assessment *for* learning and Assessment *of* learning.

The *CNPM Assessment Packs* consist of three key features:

- Section 1: Adult Directed Tasks
- Section 2: Pupil Self assessments
- Section 3: Tests

Section 1: Adult Directed Tasks

Purposes

- To assist in identifying particular children's strengths and weaknesses.
- To inform future planning and teaching of individual children and the class as a whole.
- To provide some guidance about what to do for those children who are achieving above or below expectations.

When to use this feature

- Anytime throughout the year when you are uncertain about a child's, or a group of children's, understanding of a particular objective.

Task 6

Counting and understanding number

Objective(s) for the task. Key objectives are shown in **bold**

Link to National Curriculum Attainment Target and Level

Objective **NC AT 2** **NC Level 2**

- Estimate a number of objects; round two-digit numbers to the nearest 10

List of resources including reference to the appropriate resource copymaster(s) (RCMs)

Resources

- large collection of counting objects, e.g. counters, interlocking cubes, buttons
- RCM 10: Round to 10 (per child)
- pencil (per child)
- 2 × 0–9 die (per group or pair)
- RCM 6: Three-digit number cards (for children achieving *above* expectation)
- RCM 5: Two-digit number cards (for children achieving *below* expectation)
- 1–100 number square (for children achieving *below* expectation)

Task

- This task is best undertaken with a small group of children, say 4–6, playing the game as a group. Alternatively, if you are doing this task with a larger group, arrange the children into pairs to play the game, and monitor their performance as they play them.

Success criterion

Success criterion: *Estimate a number of objects*

Questions to ask

Description of the task

- Ask a child to take a handful of counting objects and place them on the table. Ask: **How many cubes do you think there are?**
- Allow the child sufficient time to estimate the number of counting objects.
- Once the child has done this, reorganise the objects into a pile again and ask: **How many cubes are there now? Is the number of cubes still the same? How do you know?**
- Collect the counting objects and repeat the above for other children in the group.
- Continue until each child has sufficiently demonstrated their ability to estimate a group of objects.

Success criterion: *Round two-digit numbers to the nearest 10*

- Provide each child with a copy of the 'Round to 10' game and a pencil.
- Explain the rules of the game to the children:
  - Children take turns to roll two dice, e.g. 3 and 7, and use the two digits to make a two-digit number, i.e. 37 or 73.
  - They then round their two-digit number to the nearest multiple of 10, i.e. 40 or 70, and write it on the sail above the corresponding multiple of 10:

or

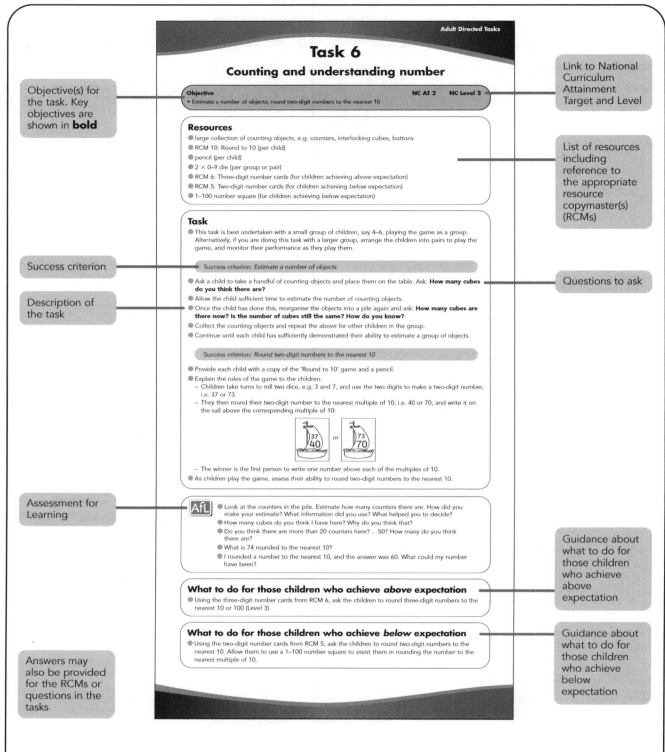

  - The winner is the first person to write one number above each of the multiples of 10.
- As children play the game, assess their ability to round two-digit numbers to the nearest 10.

Assessment for Learning

AfL

- Look at the counters in the pile. Estimate how many counters there are. How did you make your estimate? What information did you use? What helped you to decide?
- How many cubes do you think I have here? Why do you think that?
- Do you think there are more than 20 counters here? …50? How many do you think there are?
- What is 74 rounded to the nearest 10?
- I rounded a number to the nearest 10, and the answer was 60. What could my number have been?

Guidance about what to do for those children who achieve above expectation

What to do for those children who achieve *above* expectation

- Using the three-digit number cards from RCM 6, ask the children to round three-digit numbers to the nearest 10 or 100 (Level 3).

Guidance about what to do for those children who achieve below expectation

What to do for those children who achieve *below* expectation

- Using the two-digit number cards from RCM 5, ask the children to round two-digit numbers to the nearest 10. Allow them to use a 1–100 number square to assist them in rounding the number to the nearest multiple of 10.

Answers may also be provided for the RCMs or questions in the tasks

How to use this feature

- Using **Record-keeping format 1: Adult Directed Task assessment sheet**:
  - complete the top section
  - copy the Success criteria from the relevant Adult Directed Task
  - write down the names of the children you are going to be working with.

Record-keeping format 1 Adult Directed Task assessment sheet

Objective(s): *Identify reflective symmetry in patterns and 2-D shapes and draw lines of symmetry in shapes*

Date: *05/05/2008* Adult: *Lyn Davis*

NC Level: *2* Class: *2A*

Child's name	Success criteria				Other observations	Objective(s) achieved	Future action
	Identify reflective symmetry in patterns	*Identify reflective symmetry in 2-D shapes*	*Draw lines of symmetry in shapes*				
Emily Robinson							
Jerzy Novak							
Jyoti Chandra							
James Allaway							

- Use **Record-keeping format 1** to record individual children's performance during the task, commenting upon particular strengths and weaknesses, how competent you feel the children are with this objective(s) and any future action you may consider appropriate.

Section 2: Pupil Self assessments

Purpose

- To provide children with the opportunity to undertake some form of self assessment at the end of a unit.

When to use this feature

- At the end of each unit.

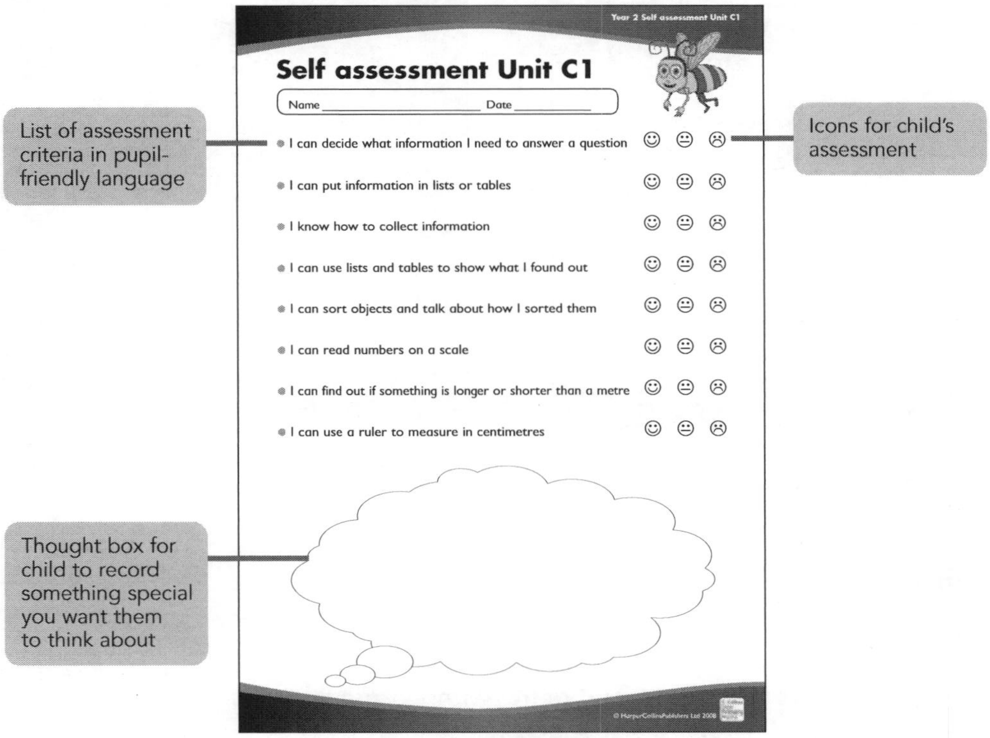
Year 2 Self assessment Unit C1

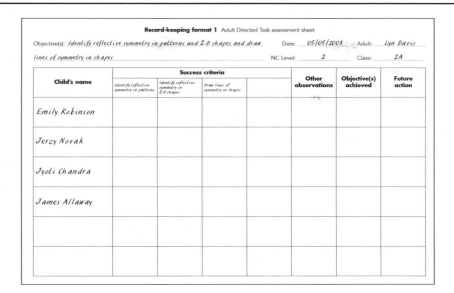

Self assessment Unit C1

Name ______ Date ______

- I can decide what information I need to answer a question
- I can put information in lists or tables
- I know how to collect information
- I can use lists and tables to show what I found out
- I can sort objects and talk about how I sorted them
- I can read numbers on a scale
- I can find out if something is longer or shorter than a metre
- I can use a ruler to measure in centimetres

© HarperCollinsPublishers Ltd 2008

How to use this feature

- Distribute the relevant Pupil Self assessment sheet at the end of the unit.
- The empty thought box at the bottom of the sheet is designed to be used by the children to record anything special that you might like them to think about, e.g.
  – anything they feel they need more practice on
  – what they think they should or could learn next
  – any special equipment that they used to help them during the unit
  – anything they particularly liked or disliked that they did during the unit.
- Ask the children to complete the sheet independently.
- After the children have completed the sheet, as a class, discuss specific objectives, asking individual children to comment on what they have written.

Section 3: Tests

Purposes

- To provide an indication of how individual children, and the class as a whole, have performed during a term, and to inform future planning.
- To inform teacher assessment when assigning an overall National Curriculum Level in mathematics.
- To inform teacher assessment when assigning a National Curriculum Level for each Attainment Target in mathematics.

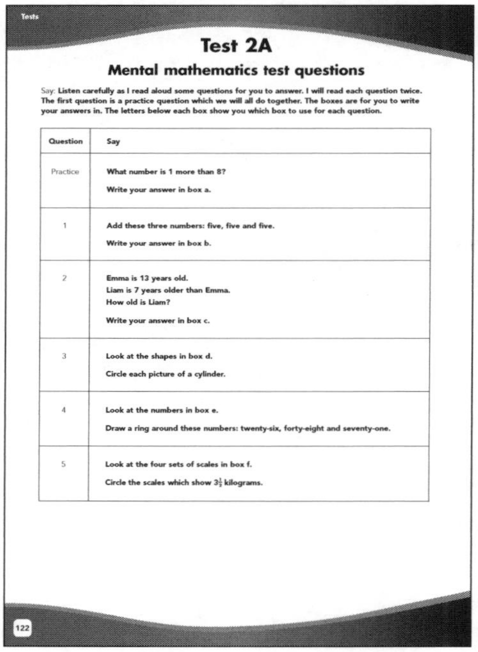
Tests

Test 2A

Mental mathematics test questions

Say: Listen carefully as I read aloud some questions for you to answer. I will read each question twice. The first question is a practice question which we will all do together. The boxes are for you to write your answers in. The letters below each box show you which box to use for each question.

Question	Say
Practice	What number is 1 more than 8? Write your answer in box a.
1	Add these three numbers: five, five and five. Write your answer in box b.
2	Emma is 13 years old. Liam is 7 years older than Emma. How old is Liam? Write your answer in box c.
3	Look at the shapes in box d. Circle each picture of a cylinder.
4	Look at the numbers in box e. Draw a ring around these numbers: twenty-six, forty-eight and seventy-one.
5	Look at the four sets of scales in box f. Circle the scales which show $3\frac{1}{2}$ kilograms.

122

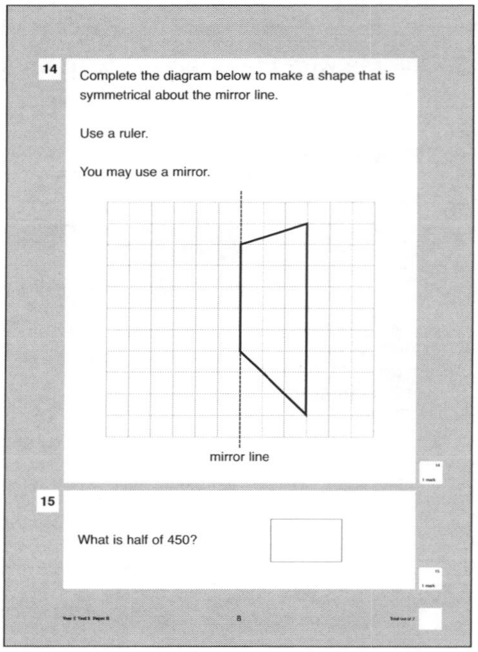

When to use this feature

- At the end of each term.

How to use this feature

- Decide which test to give each child. To assess Level 2, use Test Papers 1A, 2A and 3A. To assess Level 3, use Test Papers 1B, 2B and 3B.
- Distribute the relevant Test towards the end of the term. Ensure that all children have the necessary resources, listed on the page 1 of the test paper.

- Go through the mental practice question and the five mental mathematics questions, on pages 2 and 3 of the test paper.
- Go through the written practice question on page 4. Children then work independently to complete Paper A or B. Although there is no time limit for any part of the tests, it is recommended that you allow approximately 45 minutes for each test.
- Mark the papers and record individual children's results on their paper. You may wish to use **Record-keeping formats 2**, **3** or **4** to analyse the performance of individual children and particular test questions.

Record-keeping format 2 Test 1 grid for test analysis (Paper A: Level 2)

	Number sequence	Addition	Properties of 2-D shapes	Multiplication	Calculating time differences	Multiplication	Rounding	Identifying which operation to use	Halving: money	Reflective symmetry	Ordering whole numbers	Addition of more than two numbers	Handling data: table	Addition
AT	2	2	3	1 & 2	3	2	2	2	1 & 2	3	2	2	2	2
Question	1	2	3	4	5	6	7	8	9	10	11	12	13	14
Name / Mark	1	1	1	1	1	1	1	1	1	1	1	1	1	1
1. *James Allaway*	/	/	/	0	0	0	0	0	0	/	/	0	0	/
2. *Jyoti Chandra*	/	/	/	/	/	/	/	/	/	/	/	/	/	0
3. *Jerzy Novak*	/	/	0	/	/	0	/	/	0	/	/	0	0	/
4. *Emily Robinson*	/	/	/	/	/	/	0	/	0	/	0	0	/	0
5.														
6.														
7.														
8.														
9.														
10.														
11.														
12.														
13.														
14.														
15.														
16.														
17.														
18.														
19.														
20.														
21.														
22.														
23.														
24.														
25.														
26.														
27.														
28.														
29.														
30.														
Number correct														
Number incorrect or omitted														
Percentage correct														
Percentage incorrect or omitted														

- Using the National Curriculum Level Indicator, assign a Level for the Test.

Test Papers 1A, 2A or 3A				
National Curriculum Level Indicator				
No Level achieved	Level 1	Level 2C	Level 2B	Level 2A
0–4	5–6	7–12	13–18	19–30

Test Papers 1B, 2B or 3B	
National Curriculum Level Indicator	
Level 3 not achieved	Level 3 achieved
0–10	11–30

* These Tests must be seen only as a guide to help gaining an overall best fit in mathematics.

- Use your professional judgement of each child's overall performance during the term in each of the National Curriculum Attainment Targets. Take into account the following:
  – performance in the Test
  – observations made during Adult Directed Tasks
  – mastery of objectives from the PNS *Renewed Framework for Mathematics* (2006)
  – performance in whole class discussions
  – participation in group work
  – work presented in exercise books
  – any other written evidence.

You may wish to use the following record-keeping formats to assign a Level for each of the National Curriculum Attainment Targets:

Record-keeping format	National Curriculum Attainment Target
Record-keeping format 5	Attainment Target 1 – Using and applying mathematics
Record-keeping format 6	Attainment Target 2 – Number and algebra
Record-keeping format 7	Attainment Target 3 – Shape, space and measures

When using **Record-keeping format 5** (AT 1), once you have decided which Level best fits a particular child you may wish to identify how competent a child is at that Level by using the following key:

C	Becoming competent at this Level (Achieving up to $\frac{1}{3}$ of the Level Descriptors)	Lower
B	Competent at this Level (Achieving between $\frac{1}{3}$ and $\frac{2}{3}$ of the Level Descriptors)	Secure
A	Very competent at this Level (Achieving more than $\frac{2}{3}$ of the Level Descriptors)	Upper

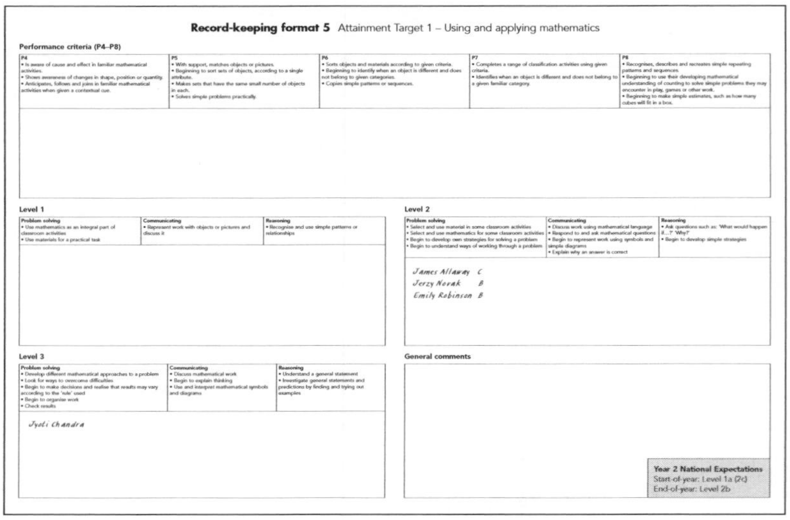
Record-keeping format 5 Attainment Target 1 – Using and applying mathematics

Performance criteria (P4–P8)

P4	P5	P6	P7	P8
• Is aware of cause and effect in familiar mathematical activities. • Shows awareness of changes in shape, position or quantity. • Anticipates, follows and joins in familiar mathematical activities when given a contextual cue.	• With support, matches objects or pictures. • Beginning to sort sets of objects, according to a single attribute. • Makes sets that have the same small number of objects in each. • Solves simple problems practically.	• Sorts objects and materials according to given criteria. • Beginning to identify when an object is different and does not belong to given categories. • Copies simple patterns or sequences.	• Completes a range of classification activities using given criteria. • Identifies when an object is different and does not belong to a given familiar category.	• Recognises, describes and recreates simple repeating patterns and sequences. • Beginning to use their developing mathematical understanding of counting to solve simple problems they may encounter in play, games or other work. • Beginning to make simple estimates, such as how many cubes will fit in a box.

Level 1

Problem solving	Communicating	Reasoning
• Use mathematics as an integral part of classroom activities • Use materials for a practical task	• Represent work with objects or pictures and discuss it	• Recognise and use simple patterns or relationships

Level 2

Problem solving	Communicating	Reasoning
• Select and use material in some classroom activities • Select and use mathematics for some classroom activities • Begin to develop own strategies for solving a problem • Begin to understand ways of working through a problem	• Discuss work using mathematical language • Respond to and ask mathematical questions • Begin to represent work using symbols and simple diagrams • Explain why an answer is correct	• Ask questions such as: 'What would happen if…?' 'Why?' • Begin to develop simple strategies

James Allaway C
Jerzy Novak B
Emily Robinson B

Level 3

Problem solving	Communicating	Reasoning
• Develop different mathematical approaches to a problem • Look for ways to overcome difficulties • Begin to make decisions and realise that results may vary according to the 'rule' used • Begin to organise work • Check results	• Discuss mathematical work • Begin to explain thinking • Use and interpret mathematical symbols and diagrams	• Understand a general statement • Investigate general statements and predictions by finding and trying out examples

Jyoti Chandra

General comments

Year 2 National Expectations
Start-of-year: Level 1a (2c)
End-of-year: Level 2b

In **Record-keeping formats 6** and **7** (AT 2 & AT 3), the Level 1 and Level 2 descriptors have been broken down into the sublevels A, B, C.

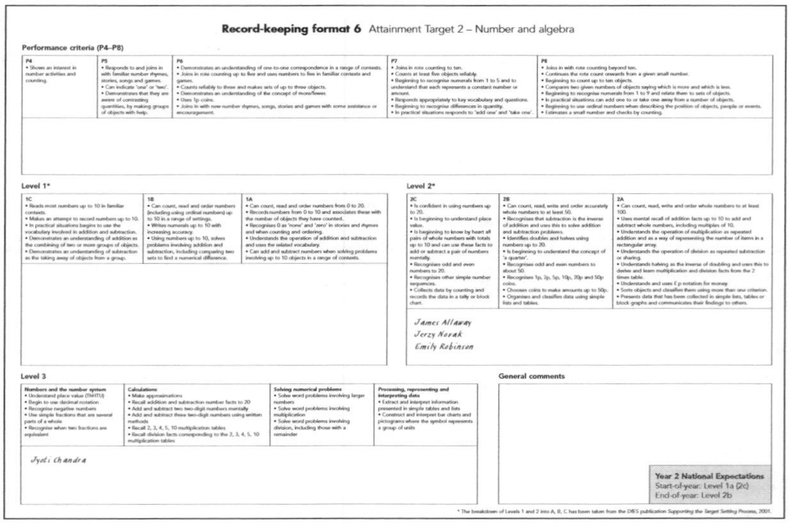
Record-keeping format 6 Attainment Target 2 – Number and algebra

Performance criteria (P4–P8)

P4	P5	P6	P7	P8
• Shows an interest in number activities and counting.	• Responds to and joins in with familiar number rhymes, stories, songs and games. • Can indicate 'one' or 'two'. • Demonstrates that they are aware of contrasting quantities, by making groups of objects with help.	• Demonstrates an understanding of one-to-one correspondence in a range of contexts. • Joins in rote counting up to five and uses numbers to five in familiar contexts and games. • Counts reliably to three and makes sets of up to three objects. • Demonstrates an understanding of the concept of more/fewer. • Uses 1p coins. • Joins in with new number rhymes, songs, stories and games with some assistance or encouragement.	• Joins in rote counting to ten. • Counts at least five objects reliably. • Beginning to recognise numerals from 1 to 5 and to understand that each represents a constant number or amount. • Responds appropriately to key vocabulary and questions. • Beginning to recognise differences in quantity. • In practical situations responds to 'add one' and 'take one'.	• Joins in with rote counting beyond ten. • Continues the rote count onwards from a given small number. • Beginning to count up to ten objects. • Compares two given numbers of objects saying which is more and which is less. • Beginning to recognise numerals from 1 to 9 and relate them to sets of objects. • In practical situations can add one to or take one away from a number of objects. • Beginning to use ordinal numbers when describing the position of objects, people or events. • Estimates a small number and checks by counting.

Level 1*

1C	1B	1A
• Reads most numbers up to 10 in familiar contexts. • Makes an attempt to record numbers up to 10. • In practical situations begins to use the vocabulary involved in addition and subtraction. • Demonstrates an understanding of addition as the combining of two or more groups of objects. • Demonstrates an understanding of subtraction as the taking away of objects from a group.	• Can count, read and order numbers (including using ordinal numbers) up to 10 in a range of settings. • Writes numerals up to 10 with increasing accuracy. • Using numbers up to 10, solves problems involving addition and subtraction, including comparing two sets to find a numerical difference.	• Can count, read and order numbers from 0 to 20. • Records numbers from 0 to 10 and associates these with the number of objects they have counted. • Recognises 0 as 'none' and 'zero' in stories and rhymes and when counting and ordering. • Understands the operation of addition and subtraction and uses the related vocabulary. • Can add and subtract numbers when solving problems involving up to 10 objects in a range of contexts.

Level 2*

2C	2B	2A
• Is confident in using numbers up to 20. • Is beginning to understand place value. • Is beginning to know by heart all pairs of whole numbers with totals up to 10 and can use these facts to add or subtract a pair of numbers mentally. • Recognises odd and even numbers to 20. • Recognises other simple number sequences. • Collects data by counting and records the data in a tally or block chart.	• Can count, read, write and order accurately whole numbers to at least 50. • Recognises that subtraction is the inverse of addition and uses this to solve addition and subtraction problems. • Identifies doubles and halves using numbers up to 20. • Is beginning to understand the concept of 'a quarter'. • Recognises odd and even numbers to about 50. • Recognises 1p, 2p, 5p, 10p, 20p and 50p coins. • Chooses coins to make amounts up to 50p. • Organises and classifies data using simple lists and tables.	• Can count, read, write and order whole numbers to at least 100. • Uses mental recall of addition facts up to 10 to add and subtract whole numbers, including multiples of 10. • Understands the operation of multiplication as repeated addition and as a way of representing the number of items in a rectangular array. • Understands the operation of division as repeated subtraction or sharing. • Understands halving as the inverse of doubling and uses this to derive and learn multiplication and division facts from the 2 times table. • Understands and uses £ p notation for money. • Sorts objects and classifies them using more than one criterion. • Presents data that has been collected in simple lists, tables or block graphs and communicates their findings to others.

James Allaway
Jerzy Novak
Emily Robinson

Level 3

Numbers and the number system	Calculations	Solving numerical problems	Processing, representing and interpreting data
• Understand place value (ThHTU) • Begin to use decimal notation • Recognise negative numbers • Use simple fractions that are several parts of a whole • Recognise when two fractions are equivalent	• Make approximations • Recall addition and subtraction number facts to 20 • Add and subtract two two-digit numbers mentally • Add and subtract three two-digit numbers using written methods • Recall 2, 3, 4, 5, 10 multiplication tables • Recall division facts corresponding to the 2, 3, 4, 5, 10 multiplication tables	• Solve word problems involving larger numbers • Solve word problems involving multiplication • Solve word problems involving division, including those with a remainder	• Extract and interpret information presented in simple tables and lists • Construct and interpret bar charts and pictograms where the symbol represents a group of units

Jyoti Chandra

General comments

Year 2 National Expectations
Start-of-year: Level 1a (2c)
End-of-year: Level 2b

* The breakdown of Levels 1 and 2 into A, B, C has been taken from the DfES publication Supporting the Target Setting Process, 2001.

Record-keeping formats 5, **6** and **7** also include the Performance Criteria (P4–P8). These Performance Criteria may be useful in identifying where those children who have not yet achieved Level 1 are working. The breakdown of AT 2 and AT 3 into the sublevels, along with P Levels, has been taken from the DfES publication *Supporting the Target Setting Process* (2001).

- You may also wish to record children's attainment in each of the Key objectives (also referred to as end-of-year expectations) using either of the following record-keeping formats:

Record-keeping format 8: Class record of the end-of-year expectations

Record-keeping format 8 Class record of the end-of-year expectations

Class: 2A

Date: 05/05/2008

Year 2 End-of-year expectations	Names: James Allaway	Jyoti Chandra	Jerzy Novay	Emily Robinson																										
Counting and understanding number (AT2) Count up to 100 objects by grouping them and counting in tens, fives or twos; explain what each digit in a two-digit number represents, including numbers where 0 is a place holder; partition two-digit numbers in different ways, including into multiples of 10 and 1 (Level 2)																														
Knowing and using number facts (AT2) Derive and recall all addition and subtraction facts for each number to at least 10, all pairs with totals to 20 and all pairs of multiples of 10 with totals up to 100 (Level 2)																														
Calculating (AT2) Add or subtract mentally a one-digit number or a multiple of 10 to or from any two-digit number; use practical and informal written methods to add and subtract two-digit numbers (Level 2)																														
Use the symbols +, −, ×, ÷ and = to record and interpret number sentences involving all four operations; calculate the value of an unknown in a number sentence (Level 2)																														
Understanding shape (AT3) Visualise common 2-D shapes and 3-D solids; identify shapes from pictures of them in different positions and orientations; sort, make and describe shapes, referring to their properties (Level 2)																														
Measuring (AT3) Use units of time (seconds, minutes, hours, days) and know the relationships between them; read the time to the quarter hour; identify time intervals, including those that cross the hour (Level 2)																														
Handling data (AT2) Use lists, tables and diagrams to sort objects; explain choices using appropriate language, including 'not' (Level 2)																														

NOTES: **Using and applying mathematics (AT1)** is incorporated throughout
End-of-year National Expectations: Level 2b

Record-keeping format 9: Individual child's record of end-of-year expectations

Record-keeping format 9 Individual child's record of the end-of-year expectations

Name: Emily Robinson

Foundation Stage		**Year 1**		**Year 2**		**Year 3**	
Using and applying mathematics (AT1)							
Use developing mathematical ideas and methods to solve practical problems (Level 1)	B						
Talk about, recognise and recreate simple patterns (Level 1)	B						
Counting and understanding number (AT2)							
Say and use the number names in order in familiar contexts (Level 1)	B	Read and write numerals from 0 to 20, then beyond; use knowledge of place value to position these numbers on a number track and number line (Level 2)	B	Count up to 100 objects by grouping them and counting in tens, fives or twos; explain what each digit in a two-digit number represents, including numbers where 0 is a place holder; partition two-digit numbers in different ways, including into multiples of 10 and 1 (Level 2)	B	Partition three-digit numbers into multiples of 100, 10 and 1 in different ways (Level 3)	
Count reliably up to 10 everyday objects (Level 1)	B						
Use language such as 'more' or 'less' to compare two numbers (Level 1)	B						
Recognise numerals 1 to 9 (Level 1)	B						
Knowing and using number facts (AT2)							
Find one more or one less than a number from 1 to 10 (Level 1)	A	Derive and recall all pairs of numbers with a total of 10 and addition facts for totals to at least 5; work out the corresponding subtraction facts (Level 2)	B	Derive and recall all addition and subtraction facts for each number to at least 10, all pairs with totals to 20 and all pairs of multiples of 10 with totals up to 100 (Level 2)	B	Derive and recall all addition and subtraction facts for each number to 20, sums and differences of multiples of 10 and number pairs that total 100 (Level 3)	
Calculating (AT2)							
Begin to relate addition to combining two groups of objects and subtraction to 'taking away' (Level 1)	B	Use the vocabulary related to addition and subtraction and symbols to describe and record addition and subtraction number sentences (Level 2)	B	Add or subtract mentally a one-digit number or a multiple of 10 to or from any two-digit number; use practical and informal written methods to add and subtract two-digit numbers (Level 2)	B	Add or subtract mentally combinations of one-digit and two-digit numbers (Level 3)	
In practical activities and discussion begin to use the vocabulary involved in adding and subtracting (Level 1)	B			Use the symbols +, −, ×, ÷ and = to record and interpret number sentences involving all four operations; calculate the value of an unknown in a number sentence (Level 2)	B		
Understanding shape (AT3)							
Use language such as 'circle' or 'bigger' to describe the shape and size of solids and flat shapes (Level 1)	B	Visualise and name common 2-D shapes and 3-D solids and describe their features; use them to make patterns, pictures and models (Level 2)	B	Visualise common 2-D shapes and 3-D solids; identify shapes from pictures of them in different positions and orientations; sort, make and describe shapes, referring to their properties (Level 2)	B	Draw and complete shapes with reflective symmetry and draw the reflection of a shape in a mirror line along one side (Level 3)	
Use everyday words to describe position (Level 1)	B						
Measuring (AT3)							
Use language such as 'greater', 'smaller', 'heavier' or 'lighter' to compare quantities (Level 1)	A	Estimate, measure, weigh and compare objects, choosing and using suitable uniform non-standard or standard units and measuring instruments, e.g. a lever balance, metre stick or measuring jug (Level 2)	B	Use units of time (seconds, minutes, hours, days) and know the relationships between them; read the time to the quarter hour; identify time intervals, including those that cross the hour (Level 2)	B	Read, to the nearest division and half-division, scales that are numbered or partially numbered; use the information to measure and draw to a suitable degree of accuracy (Level 3)	
		Handling data (AT4)					
		Answer a question by recording information in lists and tables; present outcomes using practical resources, pictures, block graphs or pictograms (Level 2)	A	Use lists, tables and diagrams to sort objects; explain choices using appropriate language, including 'not' (Level 2)	B	Use Venn diagrams or Carroll diagrams to sort data and objects using more than one criterion (Level 3)	

NOTE: **Using and applying mathematics (AT1)** is incorporated throughout

Task 1

Using and applying mathematics

Objectives **NC AT 1** **NC Level 2**

- Solve problems involving addition, subtraction, multiplication or division in contexts of numbers, measures or pounds and pence
- Identify and record the information or calculation needed to solve a puzzle or problem; carry out the steps or calculations and check the solution in the context of the problem

Resources

- RCM 1: Word problem cards (enlarged to A3 and cut out)
- pencil and paper (per child)
- appropriate apparatus such as a number line or 1–100 number square (for children achieving *below* expectation)

Task

- Give each child one of the Word problem cards from RCM 1 and a pencil and a piece of paper. See below for guidance as to which card to give to individual children depending on their ability.

	Easy	**Moderate**	**Difficult**
Cards	1–6	7–14	15–20

Success criterion: *Read and understand the problem*

- Ask the children to read the card quietly to themselves.
- In turn, go around the group asking each child to explain their word problem to the rest of the group in their own words. Ask: **Tom, what is your problem about? Vijay, what do you have to find out?**

Success criterion: *Correctly identify which operation(s) to use*

- Ask each child to suggest which operation(s) they need to use to work out the answer to the word problem. Ask: **Tom, which operation(s) do you need to use to work out the answer to your problem?**
- Ask each child to explain how they know which operation to use. Ask: **Gita, how do you know you need to add/subtract/multiply/divide? What clues are there in the problem?**

Success criterion: *Carry out the calculation(s) to obtain the correct answer using an appropriate method*

- Ask each child to write down the calculation needed to solve the problem and to work out the answer. Say: **On your sheet of paper I want each of you to write down the calculation(s) needed to solve your problem and then I want you to work out the answer.**
- After sufficient time, ask each child to read out the calculation and answer to their problem.
- Encourage each child to talk about the method they used to obtain their answer. Ask: **Vijay, how did you work that calculation out? Show us your working. Is there another way you could have worked it out?**

Success criterion: *Check answer using an effective method*

- Ask each child to check their answer. Say: **I want each of you now to check your answer.**
- After sufficient time, ask each child to explain the method they used for checking their answer.
- Repeat the above for the other word problems.
- Conclude by giving each child a calculation, e.g. 24 + 53, 38 − 5, 5 × 4, 30 ÷ 5, and asking them to make up a word problem that can be solved using the calculation. Say: **Tom, here is your calculation: 38 − 5. Tell us a word problem using this calculation.**

AfL

- How many calculations are needed to solve this problem?
- What is the first thing you need to do to work out the answer to this problem?
- Roughly, what is the answer you expect to get from this problem?
- How did you know you needed to add/subtract/multiply/divide? What clues were there in the problem?
- How are you going to record your working?
- What does this answer tell you?
- What are the important things you need to remember when solving word problems?

What to do for those children who achieve *above* expectation

- Use the grid opposite to choose suitable word problem cards.

What to do for those children who achieve *below* expectation

- Use the grid opposite to choose suitable word problem cards.
- Allow the children to use appropriate apparatus to work out the answer to the calculation, e.g. a number line or 1–100 number square.

Answers

Question	Problem involving	Number of steps required Operation(s) required	Answer	Question	Problem involving	Number of steps required Operation(s) required	Answer
1.	Real life	1 step: addition	20	11.	Real life	1 step: subtraction	20
2.	Money	1 step: halving	50p	12.	Real life	1 step: subtraction	14
3.	Real life	1 step: addition	61	13.	Real life	1 step: multiplication	80
4.	Measures: Length	1 step: halving	15 cm	14.	Real life	1 step: multiplication	30
5.	Measures: Capacity	1 step: addition	8 litres	15.	Real life	2 steps: addition and multiplication	60
6.	Real life	1 step: addition	23	16.	Real life	1 step: multiplication or doubling	14
7.	Real life	1 step: division	2	17.	Money	2 steps: multiplication or doubling, and subtraction	£8
8.	Measures: Time	1 step: addition or subtraction	15 minutes or $\frac{1}{4}$ hour	18.	Real life	2 steps: subtraction twice, or addition and subtraction	7
9.	Measures: Weight	1 step: subtraction	13 kg	19.	Real life	1 step: subtraction	13
10.	Real life	1 step: subtraction	23	20.	Real life	2 steps: doubling and addition	21

Using and applying mathematics

Objective **NC AT 1** **NC Level 2**

- Follow a line of enquiry; answer questions by choosing and using suitable equipment and selecting, organising and presenting information in lists, tables and simple diagrams

See Task 23
Handling data and Using and applying mathematics
Page 67

Task 2

Using and applying mathematics

Objectives **NC AT 1** **NC Level 2**

- Describe patterns and relationships involving numbers or shapes, make predictions and test these with examples
- Present solutions to puzzles and problems in an organised way; explain decisions, methods and results in pictorial, spoken or written form, using mathematical language and number sentences
- Identify and record the information or calculation needed to solve a puzzle or problem; carry out the steps or calculations and check the solution in the context of the problem

Resources

- RCM 2: Puzzles 1
- RCM 3: Puzzles 2
- ruler (per child)
- pencil and paper (per child)

Task

- Prior to the task, decide which puzzle/investigation to give individual children from RCM 2 and RCM 3. Alternatively, use puzzles or investigations of your own.

Success criteria: *Describe patterns and relationships*
Make predictions
Solve puzzles and problems
Explain decisions, methods and results

- Give each child a puzzle/investigation from RCM 2 or RCM 3 and a pencil and a piece of paper.
- Ask each child to read through their puzzle/investigation. Ask: **Andy, what is your puzzle/investigation about? What do you have to find out? What do you know already that can help you solve this?**
- Briefly discuss the puzzle/investigation with each child.
- Say: **I now want each of you to work on your puzzle/investigation. If you need anything, or are unsure of something just ask me.**
- Allow the children sufficient time to spend on their puzzle/investigation. As the children work through the task, ask specific questions to help individual children with the task as well as to assess children's ability to interpret and complete the task.
- Once each child has completed their task, ask each child to talk about what they did and what they found out.
- Ask each child to justify why they worked the way they did. Encourage them to explain their methods of working and recording. Ask: **Why did you...? How else could you have gone about it? What did you find easy/difficult about what you did? If you had to do this puzzle/investigation again, how would you do it differently next time?**
- If appropriate, repeat the above for other puzzles/investigations.

- What do you have to do in this puzzle/investigation?
- Explain to me what you have discovered.
- Write about what you discovered. Can you explain this using diagrams and symbols instead of words?

What to do for those children who achieve *above* expectation

- Provide more challenging puzzles/investigations.

What to do for those children who achieve *below* expectation

- Provide extra support as the children work through the puzzle/investigation.
- Choose easier puzzles/investigations.

Answers

RCM 2: Puzzles 1

1.

2.

3. Tracey is right both times.

RCM 3: Puzzles 2

4.

6	7	5
5	**6**	**7**
7	**5**	6

5.

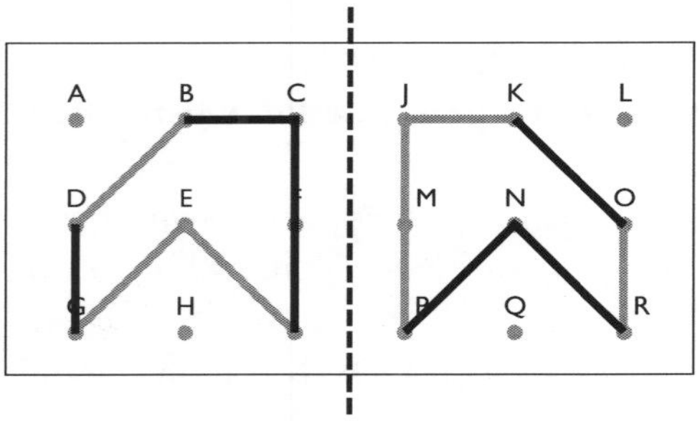

6. 22 = 15 + 7 or 49 − 27
16 = 34 − 18 or 56 − 40
33 = 18 + 15 or 40 − 7
71 = 15 + 56 or 89 − 18
83 = 49 + 34 or 56 + 27
7 = 56 − 49 or 34 − 27
Other calculations are possible.

Task 3

Counting and understanding number

Objective **NC AT 2** **NC Level 2**

- Read and write two-digit and three-digit numbers in figures and words; describe and extend number sequences and recognise odd and even numbers

Resources

- RCM 4: 1–20 number cards
- RCM 5: Two-digit number cards
- RCM 6: Three-digit number cards
- large sheet of paper and marker
- RCM 8: Continuing number sequences (enlarged to A3)
- pencil and paper (per child)
- RCM 7: Four-digit number cards (for children achieving *above* expectation)

Task

NOTE: This task involves a number of Success criteria. It is advisable to choose just one or two at a time.

- Prior to the task, on the large sheet of paper, write several two-digit and three-digit numbers in words, e.g. fifteen, seventy-nine, thirty-four, sixty, one hundred and eighteen, three hundred and seventy, six hundred and ninety-two…
- Shuffle the one-digit, two-digit and three-digit number cards from RCMs 4, 5 and 6 and spread them out face up on the table.
- Provide each child with a pencil and a piece of paper.

Success criterion: *Read two-digit and three-digit numbers in figures*

- Point to specific numbers and ask: **Fina, what is this number? Mike, tell me this number.**
- Ask individual children to point to specific numbers, e.g. **Josie, point to the number one hundred and thirteen. Fina point to seventy-nine.**
- Repeat the above until each child has sufficiently demonstrated their ability to read two-digit and three-digit numbers in figures.

Success criterion: *Read two-digit and three-digit numbers in words*

- Referring to the large sheet of paper with the two-digit and three-digit numbers written in words, point to one of the numbers and ask: **Josie, what is this number? Fina, tell me this number.**
- Ask individual children to point to specific numbers, e.g. **Mike, point to the number one hundred and eighteen. Fina, point to thirty-four.**
- Repeat the above until each child has sufficiently demonstrated their ability to read two-digit and three-digit numbers in words.

Success criterion: *Write two-digit and three-digit numbers in figures*

- Say a specific two-digit or three-digit number, e.g. **42**, and ask a child to write down the number as a figure.
- Repeat several times for each child.
- Ask each child to tell you a different two-digit or three-digit number they know. Ask: **Mike, can you tell me a different two-digit number?**
- Ask each child to write down the two-digit number they suggest as a figure.
- Repeat the above until each child has sufficiently demonstrated their ability to write two-digit and three-digit numbers in figures.

Success criterion: *Write two-digit and three-digit numbers in words*

- Referring back to the number cards from RCMs 4, 5 and 6, point to a specific number card and ask a child to write down the number as a word.
- Repeat several times for each child.
- Ask each child to tell you a two-digit or three-digit number that is different from those on the number cards. Ask: **Josie, can you tell me a three-digit number that is different from those on the cards?**
- Ask each child to write down the two-digit or three-digit number they suggest as a word.
- Repeat the above until each child has sufficiently demonstrated their ability to write two-digit and three-digit numbers in words.

Success criterion: *Describe and extend number sequences*

- Show the children RCM 8: Continuing number sequences.
- Referring to the first sequence, say: **Look at this number sequence**. Ask: **What is the rule in my sequence?** (+ 2) **Fina, continue this number sequence for me.**
- Repeat for the remaining number sequences on RCM 8.
- Occasionally, alternate this by turning over one of the number cards from RCMs 4, 5 and 6 and saying the number to the children, e.g. 13.
- Count on or back in steps, e.g. say: **13, 15, 17, 19, 21, 23**.
- Ask: **What is the rule in my sequence?** (+ 2) **Mike, continue this number sequence for me.**
- Repeat the above until each child has sufficiently demonstrated their ability to describe and extend number sequences.

Success criterion: *Count on in steps*

- Referring back to the number cards from RCMs 4, 5 and 6, point to one of the number cards and ask: **Josie, what is this number?**
- Say: **I want you to count on in steps of five from this number until I say stop. Ready? Go!**
- Continue until the child has continued the count for at least six steps.
- Repeat the above until each child has sufficiently demonstrated their ability to count on in steps of 1, 2, 3, 4, 5 and 10.

Success criterion: *Count back in steps*

- Once again point to one of the number cards and ask: **Mike, what is this number?**
- Say: **I want you to count back in steps of two from this number until I say stop. Ready? Go!**
- Continue until the child has continued the count for at least six steps.
- Repeat the above until each child has sufficiently demonstrated their ability to count back in steps of 1, 2, 3, 4, 5 and 10.

Success criterion: *Recognise and suggest odd and even numbers*

- Still referring to the number cards from RCMs 4, 5 and 6, point to a specific number card and ask: **Fina, is this an odd or an even number?**
- Ask: **Can you tell me another odd/even number?**
- Repeat the above until each child has sufficiently demonstrated they can recognise and suggest odd and even numbers.

AfL

- What is this number?
- Write the number thirty-eight for me in figures. Can you write it for me in words also?
- Which of these numbers is easiest to write down: 231, 201, 230? Why?
- Count on from zero in steps of 5 until I stay stop. Now count back to zero in steps of 5.
- What is the next number in this sequence? How do you know this is the next number?
- What is the rule for this sequence?
- Is this an odd or an even number? How do you know?
- Tell me an odd number. …even number.

What to do for those children who achieve *above* expectation

- Include the four-digit number cards from RCM 7 and ask the children to read and write whole numbers to 10 000. (Level 3)
- Ask the children to count on and back in any single-digit step and in multiples of 10. (Level 3)

What to do for those children who achieve *below* expectation

- Only use the number cards from RCM 4 and RCM 5 and ask the children to read and write numbers to 100.
- Only ask the children to count on or back in steps of 1, 2, 5 and 10.
- Only ask the children to identify odd and even numbers to 20.

Task 4

Counting and understanding number

Objective **NC AT 2** **NC Level 2**

- **Count up to 100 objects by grouping them and counting in tens, fives or twos; explain what each digit in a two-digit number represents, including numbers where 0 is a place holder; partition two-digit numbers in different ways, including into multiples of 10 and 1**

Resources

- large collection of counting objects, e.g. counters, interlocking cubes, buttons
- RCM 4: 1–20 number cards
- RCM 5: Two-digit number cards
- large sheet of paper and marker
- pencil and paper (per child and yourself)
- RCM 6: Three-digit number cards (for children achieving *above* expectation)

Task

- Shuffle the 10–20 number cards from RCM 4 and the two-digit number cards from RCM 5 and place them in a pile.

Success criterion: *Count up to 100 objects by grouping them and counting in tens, fives or twos*

- Ask a child to take a handful of counting objects and place them on the table. Ask: **How many cubes do you think there are? How could you find out?**
- Allow the child sufficient time to count the counting objects.
- Once the child has done this, reorganise the objects into a pile again and ask: **How many cubes are there now? Is the number of cubes still the same? How do you know? How can you count these cubes in a different way?** If the child does not suggest counting them in tens, fives or twos, suggest this to the child and encourage them to count the objects using this method.
- Collect the counting objects and repeat the above for other children in the group.
- Continue until each child has sufficiently demonstrated their ability to count up to 100 objects by grouping them and counting in tens, fives or twos.

Success criterion: *Identify errors in counting up to 100 objects*

- Collect a handful of counting objects, e.g. 14, and place them in front of the children. Say: **I'm going to count how many counters there are, but you need to listen carefully in case I make a mistake. Ready? 1, 2, 3, 4, 5, 6, 7, 8, 10, 11, 12, 13, 14, 15**.
- When you have finished counting say: **15 counters**. Ask: **Is this right? Why not?**
- Recount the counters correctly.
- Repeat the above, making the following mistakes when counting:
  – omitting numbers
  – repeating numbers
  – using the wrong number names
  – over-counting
  – under-counting.

Success criterion: *Explain what each digit in a two-digit number represents, including numbers where 0 is a place holder*

- Place a two-digit number card from RCM 4 or RCM 5 face up in front of each child.
- Point to a specific digit on the number card and ask: **Ben, what does this digit represent?**
- Repeat several times for each child.
- Place a different two-digit number card face up in front of each child and say: **Jason, point to the digit that shows how many units/tens are in this number.**
- Repeat the above until each child has sufficiently demonstrated their ability to explain what each digit in a two-digit number represents, including numbers where 0 is a place holder.

Success criterion: *Partition two-digit numbers into multiples of 10 and 1*

- On the large sheet of paper, write a two-digit number in expanded notation, e.g. 70 + 9, ask: **Ben, what is this number? Can you write it for me as a two-digit number?**
- Repeat several times for the other children in the group.
- Provide each child, and yourself, with a pencil and a piece of paper.
- Place a two-digit number card from RCM 5 face up in front of each child and yourself.
- Referring to the number in front of you, e.g. 57, write and say: **I can partition the number 57 into 50 plus 7.** (57 = 50 + 7)
- Say: **I want each of you to look at the number card in front of you and partition it into tens and units.**
- Repeat the above until each child has sufficiently demonstrated their ability to partition two-digit numbers into multiples of 10 and 1.

Success criterion: *Partition two-digit numbers in different ways*

- Referring to the number you have previously partitioned into tens and units, e.g. 57, say: **I can partition this number in a different way**. Write and say: **I can partition the number 57 into 40 + 17.** (57 = 40 + 17)
- Say: **I want each of you to look at the number card in front of you and partition it in a different way**.
- Once again, referring to the number you have previously partitioned, say: **I can partition this number into in another different way**. Write and say: **I can partition the number 57 into 30 + 27.** (57 = 30 + 27)
- Say: **I want each of you to look at the number card in front of you and partition it in another way**.
- Once the children have done this, ask: **See if you can partition your number in yet another way?**
- Repeat the above until each child has sufficiently demonstrated their ability to partition two-digit numbers in different ways.

AfL

- Tell me how many counters are in this pile. Can you find a quicker way than counting in ones?
- There are more than 30 counters here. Find out how many there are. Is there a better way than counting in twos? Why is this better than counting in ones or twos?
- What does this digit represent?
- How many tens are there in 37? How many units are there?
- Which digit shows how many units/tens are in this number?
- There are 3 tens in 30. How many tens are there in 36?
- What makes 40 and 47 different?
- What numbers can you make using the digits 7 and 2?
- *(Show the number cards for 14 and 41.)* Which of these numbers is fourteen? How do you know? What does the other one say? How are they the same/different?
- How many tens are there in 80? Use this fact to partition the number 85. Show me two other ways you might partition this number.
- Partition this number into tens and units. Can you partition this number in a different way?
- A number is partitioned like this: 70 + 4. What is the number? Can you partition it in another way?

What to do for those children who achieve *above* expectation

- Using the number cards from RCMs 4, 5 and 6, ask the children to explain what each digit in a number to at least 1000 represents and to partition three-digit numbers into multiples of 100, 10 and 1 in different ways (Level 3).

What to do for those children who achieve *below* expectation

- Ask the children to count up to 20 objects.
- Using the 'teen' number cards from RCM 4, ask the children to explain what each digit in a 'teen' number represents and to partition 'teen' numbers into multiples of 10 and 1.

Task 5

Counting and understanding number

Objective **NC AT 2** **NC Level 2**

- Order two-digit numbers and position them on a number line; use the greater than (>) and less than (<) signs

Resources

- RCM 4: 1–20 number cards
- RCM 5: Two-digit number cards
- pencil and paper (per child)
- RCM 9: Symbol cards
- RCM 6: Three-digit number cards (for children achieving *above* expectation)

Task

- Shuffle the 1–20 number cards from RCM 4 and the two-digit number cards from RCM 5 and place them in a pile.

Success criterion: *Order two-digit numbers*

- Lay five number cards, face up, in front of each child. Say: **Look at the numbers in front of you. I want each of you to place these cards in order, smallest to largest.**
- Once each child has done this, give each child another card and say: **Look at the cards you have just put in order. Where would you put this number so that the order is still correct?**
- Repeat the above until each child has sufficiently demonstrated their ability to order numbers to 100.

Success criterion: *Position two-digit numbers on a number line*

- Collect and shuffle all the number cards.
- Provide each child with a pencil and a piece of paper.
- Lay five cards, face up, in front of each child. Say: **Look at the numbers in front of you. This time I want each of you to draw an empty number line and mark the numbers in front of you on the line.**
- Repeat the above until each child has sufficiently demonstrated their ability to position numbers to 100 on a number line.

Success criterion: *Use the greater than (>) and less than (<) signs – symbol missing*

- Give each child a 'greater than / less than' card from RCM 9. Ensure that the children realise that the 'greater than / less than' card can be used to represent either symbol by turning the card upside down.
- Keep the missing number '□' cards to one side. (You won't need the equals sign cards for this task.)
- Choose two number cards from RCMs 4 and 5 and a symbol card. Place them in front of a child, leaving a space between the two cards, e.g. 41   69.
- Repeat this for each child in the group.
- Randomly spread the remaining number cards face up on the table.
- Say: **Look at the number cards in front of you. I want you to place your symbol card between these two number cards so that it makes a correct statement.**

- When the children have done this, ask each child to say their statement, e.g. 41 < 69. **41 is less than 69.**
- Collect all the cards and repeat the above until each child has sufficiently demonstrated their ability to use the greater than and less than signs correctly – symbol missing.

Success criterion: *Use the greater than (>) and less than (<) signs – number missing*

- Choose a number card from RCM 4 and RCM 5 and a symbol card and place them in front of a child, e.g. 52 <.
- Repeat this for each child in the group.
- Say: **Look at the cards in front of you. This time I want you to choose a card from the table and place it after the symbol card so that the statement is correct.**
- When the children have done this, ask each child to say their statement, e.g. 52 < 81. **52 is less than 81.**
- Collect all the cards and repeat the above until each child has sufficiently demonstrated their ability to use the greater than and less than signs correctly – number missing.

AfL

- Order this set of numbers, staring with the smallest.
- When you order a set of numbers what do you look at first?
- How do you find the smallest/largest number? What do you look for?
- What do you do when the numbers you are ordering have the same tens digit?
- Look at this expression: □ < 43. What could the missing number be? What other number could it be?
- 45 > 75. Is this statement true or false? How do you know?

What to do for those children who achieve *above* expectation

- Using the number cards from RCMs 4, 5 and 6, ask the children to order numbers to at least 1000, position them on a number line and use the greater than (>) and less than (<) signs (Level 3).

What to do for those children who achieve *below* expectation

- Only ask the children to order numbers from 1 to 20.

Task 6

Counting and understanding number

Objective **NC AT 2** **NC Level 2**

• Estimate a number of objects; round two-digit numbers to the nearest 10

Resources

- large collection of counting objects, e.g. counters, interlocking cubes, buttons
- RCM 10: Round to 10 (per child)
- pencil (per child)
- 2 × 0–9 die (per group or pair)
- RCM 6: Three-digit number cards (for children achieving *above* expectation)
- RCM 5: Two-digit number cards (for children achieving *below* expectation)
- 1–100 number square (for children achieving *below* expectation)

Task

- This task is best undertaken with a small group of children, say 4–6, playing the game as a group. Alternatively, if you are doing this task with a larger group, arrange the children into pairs to play the game, and monitor their performance as they play them.

Success criterion: *Estimate a number of objects*

- Ask a child to take a handful of counting objects and place them on the table. Ask: **How many cubes do you think there are?**
- Allow the child sufficient time to estimate the number of counting objects.
- Once the child has done this, reorganise the objects into a pile again and ask: **How many cubes are there now? Is the number of cubes still the same? How do you know?**
- Collect the counting objects and repeat the above for other children in the group.
- Continue until each child has sufficiently demonstrated their ability to estimate a group of objects.

Success criterion: *Round two-digit numbers to the nearest 10*

- Provide each child with a copy of the 'Round to 10' game and a pencil.
- Explain the rules of the game to the children:
  - Children take turns to roll two dice, e.g. 3 and 7, and use the two digits to make a two-digit number, i.e. 37 or 73.
  - They then round their two-digit number to the nearest multiple of 10, i.e. 40 or 70, and write it on the sail above the corresponding multiple of 10:

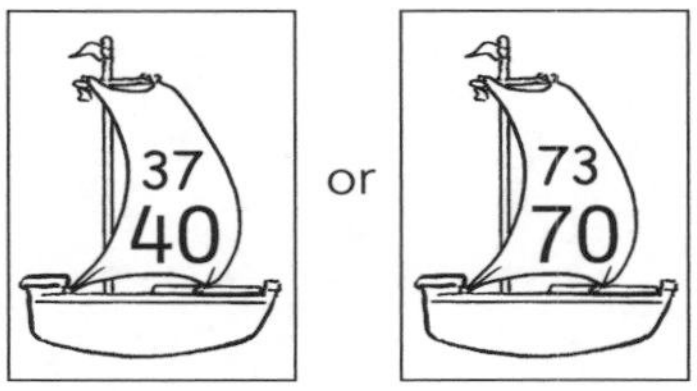

  - The winner is the first person to write one number above each of the multiples of 10.
- As children play the game, assess their ability to round two-digit numbers to the nearest 10.

AfL

- Look at the counters in the pile. Estimate how many counters there are. How did you make your estimate? What information did you use? What helped you to decide?
- How many cubes do you think I have here? Why do you think that?
- Do you think there are more than 20 counters here? ...50? How many do you think there are?
- What is 74 rounded to the nearest 10?
- I rounded a number to the nearest 10, and the answer was 60. What could my number have been?

What to do for those children who achieve *above* expectation

- Using the three-digit number cards from RCM 6, ask the children to round three-digit numbers to the nearest 10 or 100 (Level 3).

What to do for those children who achieve *below* expectation

- Using the two-digit number cards from RCM 5, ask the children to round two-digit numbers to the nearest 10. Allow them to use a 1–100 number square to assist them in rounding the number to the nearest multiple of 10.

Task 7

Counting and understanding number

Objective **NC AT 2** **NC Level 2**

- Find one half, one quarter and three quarters of shapes and sets of objects

Resources

- RCM 11: Fractions of shapes cards (enlarged to A3)
- RCM 12: Array cards (cards A–I only)
- coloured pencil (per child)
- large sheet of paper and marker (for children achieving *above* expectation)

Task

Success criterion: *Recognise one half, one quarter and three quarters of shapes*

- Lay the A–J cards from RCM 11 face up on the table.
- Ask individual children to identify a shape that shows one half, one quarter or three quarters shaded. Ask: **Joshua, can you point to a shape that shows one half shaded? Luke, can you point to a shape that is three quarters shaded?**
- Refer to the table below when choosing which card to give individual children.

	Easy	**Difficult**
Cards	A–E	F–J

- Continue until each child has sufficiently demonstrated their ability to recognise one half, one quarter and three quarters of shapes.

Success criterion: *Find one half, one quarter and three quarters of shapes*

- Give each child one of the K–T shape cards from RCM 11 and a coloured pencil.
- Ask them to shade one half, one quarter or three quarters of the shape. Say: **Joshua, can you colour one quarter of your shape? Luke, I want you to colour three quarters of your shape.**
- Refer to the table below when choosing which card to give individual children and what fraction of the shape you ask them to shade.

Level of difficulty	Easy				
Shape card	K	L	M	N	O
Fraction of shape than can be shaded	$\frac{1}{2}$ only	$\frac{1}{2}$ only	$\frac{1}{2}$, $\frac{1}{4}$ or $\frac{3}{4}$	$\frac{1}{2}$, $\frac{1}{4}$ or $\frac{3}{4}$	$\frac{1}{2}$, $\frac{1}{4}$ or $\frac{3}{4}$
Level of difficulty	Difficult				
Shape card	P	Q	R	S	T
Fraction of shape than can be shaded	$\frac{1}{2}$, $\frac{1}{4}$ or $\frac{3}{4}$	$\frac{1}{2}$, $\frac{1}{4}$ or $\frac{3}{4}$	$\frac{1}{2}$, $\frac{1}{4}$ or $\frac{3}{4}$	$\frac{1}{2}$, $\frac{1}{4}$ or $\frac{3}{4}$	$\frac{1}{2}$, $\frac{1}{4}$ or $\frac{3}{4}$

- Continue until each child has sufficiently demonstrated their ability to find one half, one quarter and three quarters of shapes.

Success criterion: *Find one half, one quarter and three quarters of sets of objects*

- Give each child one of the array cards A–I from RCM 12.
- Ask them to find one half, one quarter or three quarters of the dots on the card. Ask: **Luke, how many dots are there on this card? What is one quarter of this amount? Joshua, how many dots are there on your card? Tell me what is one quarter of this number.**
- Refer to the table below when choosing which card to give individual children and what fraction of objects to identify.

Array card	A	B	C	D	E	F	G	H	I
Fraction that dots can be divided equally into	$\frac{1}{2}$, $\frac{1}{4}$ or $\frac{3}{4}$					$\frac{1}{2}$ only			$\frac{1}{2}$, $\frac{1}{4}$ or $\frac{3}{4}$

- Continue until each child has sufficiently demonstrated their ability to find one half, one quarter and three quarters of sets of objects.

AfL

- Explain how we could find one quarter of this set of 12 crayons? What about three quarters?
- How could we give someone half of 20p if we had one 20p coin? What about half of 12p if we had two 5p and two 1p coins?
- How could we work out half of three equal strips of paper?
- How could you find one quarter of a piece of string?
- What about a quarter of two pieces of string?
- Shade one quarter of this shape.
- How will you find one quarter of that rectangle? …three quarters…?
- If one quarter of a set of pencils is 2, how many pencils are in the set?
- Here is a pizza cut into eight equal pieces. How many pieces are needed for three quarters of the pizza?

What to do for those children who achieve *above* expectation

- Write the following on the large sheet of paper:

4		8		10
	18		12	
20		24		14
	6		16	
	28		40	

- Pointing to each number in turn, ask the children to find one half, one quarter and/or three quarters of the number. Pointing to 12, ask: **Joshua, what is half of 12? What is one quarter of 12? Can you tell me what three quarters of 12 is?** Pointing to 14, ask: **Luke, what is half of 14?**

What to do for those children who achieve *below* expectation

- Only ask the children to recognise and find half of a shape divided into two equal parts and one quarter of a shape divided into four equal parts.
- Only ask the children to find one half and one quarter of a set of objects.

Task 8

Knowing and using number facts

Objective **NC AT 2** **NC Level 2**

- **Derive and recall all addition and subtraction facts for each number to at least 10, all pairs with totals to 20 and all pairs of multiples of 10 with totals up to 100**

Resources

- RCM 4: 1–20 number cards
- RCM 13: Addition facts to 10 (per child)
- pencil (per child)
- 1–10 or 0–9 die
- RCM 14: Subtraction facts to 10 (per child)
- RCM 15: Multiples of 10 number cards
- 1–6 die (for children achieving *below* expectation)
- 0–10 number line (for children achieving *below* expectation)
- 0–20 number line (for children achieving *below* expectation)
- number line marked in multiples of 10 from 0 to 100 (for children achieving *below* expectation)

Task

- Prior to the task, photocopy RCM 13 and RCM 14 back to back. Having only one sheet of paper will make the task easier for the children.
- Provide each child with a copy of RCMs 13 and 14 and a pencil.
- Explain to the children that RCM 13 has addition facts to 10 written on balloons and that RCM 14 has subtraction facts to 10 written on beach balls.
- Show the children the die.
- If you are using a 0–9 die, explain to the children that if you roll a zero on the die that it represents ten not zero.

Success criterion: *Derive and recall all addition facts for each number to 10*

- Say: **Look at the addition calculations written on the balloons. I'm going to roll this die and call out the number rolled. You then have to find just one of the addition calculations on your sheet that has that number as the answer, and write that number in the box. Let's try one.**
- Roll the die, e.g. 6.
- Say: **6. Find just one addition calculation on your sheet where the answer is 6 and write 6 in the box.**
- Quickly check that all the children understand what they are required to do.
- Repeat the above several times, quickening the pace as children become more confident with the task.

Success criterion: *Derive and recall all subtraction facts for each number to 10*

- Say: **Turn over your sheet. Look at the subtraction calculations written on the beach balls. I'm going to roll the die, and again I'm going to call out the number rolled. This time you have to find just one subtraction calculation on your sheet that has that number as the answer, and write that number in the box. Let's try one.**

- Roll the die, e.g. 4.
- Say: **4. Find one subtraction calculation on your sheet where the answer is 4 and write 4 in the box.**
- Quickly check that all the children understand what they are required to do.
- Repeat the above several times, quickening the pace as children become more confident with the task.

NOTE:

- Be aware of those numbers that keep recurring when you are rolling the die, as children may run out of calculations with that answer on their sheet.
- You may wish to concentrate solely on addition facts to 10 before moving on to the subtraction facts, or interchange between addition and subtraction facts. If you decide to change between addition and subtraction facts, be sure to tell the children which type of calculation they are to look out for.

Success criterion: *Derive and recall all pairs of numbers that total 20*

- Randomly spread the 1–20 number cards from RCM 4 face up on the table.
- Say: **These are all the numbers from 1 to 20. I'm going to place one card in front of each of you, and I want you to tell me what number you need to add to your number to make 20.**
- Demonstrate this to the children by turning over the top card, saying the number, e.g. 7, and the number that added to 7 totals 20, i.e. 13. Say: **7 plus 13 equals 20.**
- Continue until each child has sufficiently demonstrated their ability to derive and recall all pairs of numbers that total 20.

Success criterion: *Derive and recall all pairs of multiples of 10 that total 100*

- Collect the 1–20 number cards and place them to one side. Randomly spread the multiples of 10 number cards from RCM 15 face up on the table.
- Say: **I am going to give each of you one of these multiples of 10 number cards. I then want each of you to find the other multiple of 10 card that when added to your number makes 100.**
- Demonstrate this to the children by taking a card, e.g. 30, and saying: **30 add 70 equals 100**, then taking the 70 number card.
- Occasionally ask questions such as: **How did you work out that answer? What number fact to 10 helped you to work out the answer to this calculation?**
- Continue until each child has sufficiently demonstrated their ability to derive and recall all pairs of multiples of 10 that total 100.

AfL

- Tell me two numbers that have a total of 7.
- Tell me two numbers with a difference of 3.
- Tell me some other words that mean the same as 'add' / 'take away'.
- Seven and what other number total 20?
- Tell me two numbers that added together equal 20.
- 40 add what make 100?
- Can you tell me another pair of multiples of 10 that total 100?

What to do for those children who achieve *above* expectation

- Ask the children quick-fire questions involving addition and subtraction number facts to 10, including those that require the children to calculate the value of an unknown in a number sentence, e.g. 4 + 5 = □, 6 + □ = 8, □ − 6 = 3, 3 = 7 − □, 10 = 5 + □.
- Extend this to asking quick-fire questions involving addition and subtraction number facts to 20, including those that require the children to calculate the value of an unknown in a number sentence, e.g. 7 + 6 = □, 13 + □ = 19, □ − 8 = 9, 5 = 13 − □, 14 = 8 + □. (Level 3)
- Ask the children to derive sums and differences of pairs of multiples of 10, e.g. 30 + 40 = □, 50 + 60 = □, 70 − 30 = □, 90 − 50 = □. (Level 3)

What to do for those children who achieve *below* expectation

- Use a 1–6 die rather than a 1–10 or 0–9 die. This way, children will only have to answer calculations involving addition and subtraction facts to 6.
- When being asked to derive and recall all addition and subtraction facts for each number to 10, allow the children to use a 0–10 number line.
- When being asked to derive and recall all pairs of numbers that total 20, allow the children to use a 0–20 number line.
- When being asked to derive and recall all pairs of multiples of 10 that total 100, allow the children to use a number line marked in multiples of 10 from 0 to 100.

Answers

RCM 13: Addition facts to 10

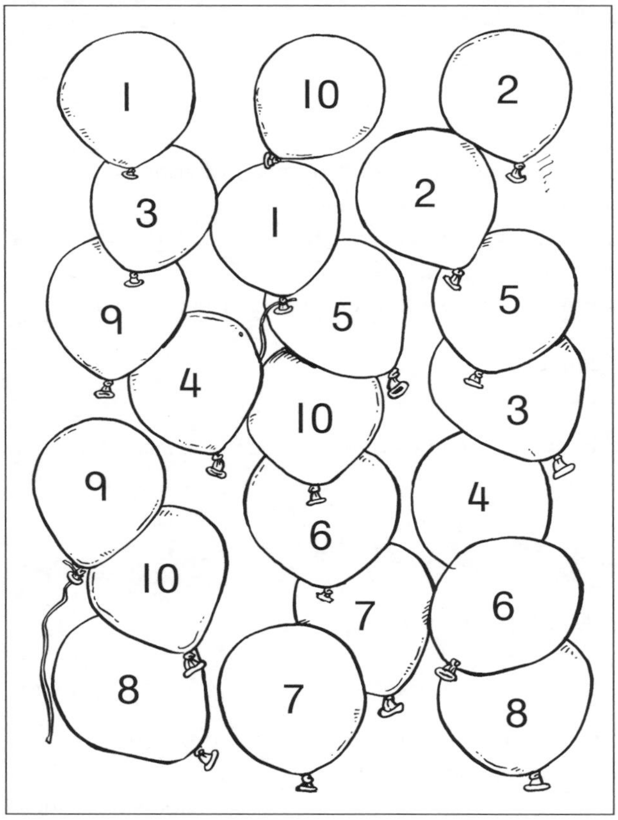

RCM 14: Subtraction facts to 10

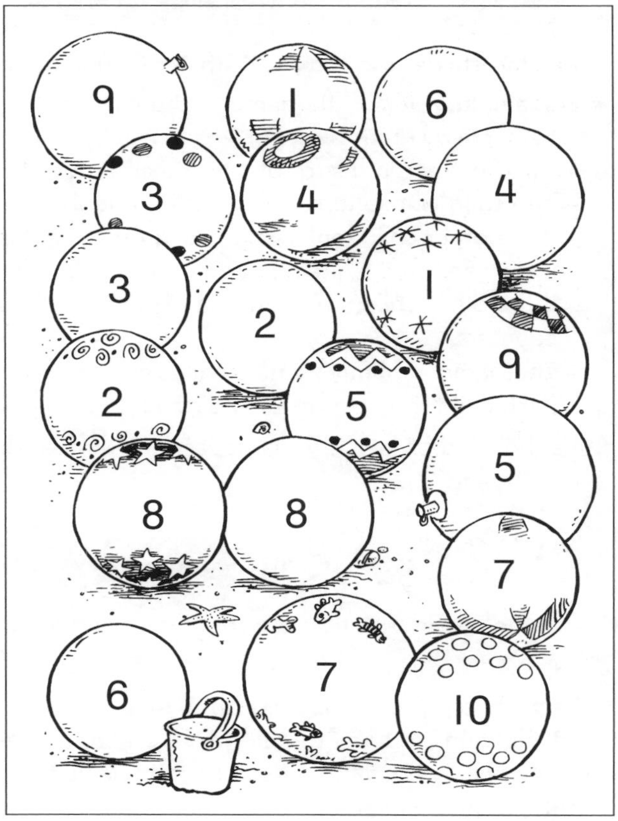

Task 9

Knowing and using number facts

Objective **NC AT 2** **NC Level 2**

- Understand that halving is the inverse of doubling and derive and recall doubles of all numbers to 20, and the corresponding halves

Resources

- RCM 16: Doubling game (per pair or group) (enlarged to A3)
- RCM 17: Halving game (per pair or group) (enlarged to A3)
- 1–6 die (per pair or group)
- container of counters (per pair or group)
- different coloured cube (per child)

Task

- Prior to the task, photocopy RCM 16 and RCM 17 back to back. Having only one sheet of paper will make the task easier for the children.
- Arrange the children into pairs or groups.
- Provide each pair/group with a copy of RCMs 16 and 17, a 1–6 die and a container of counters, and each child with a different coloured cube.
- Explain to the children that they are going to play two games: Doubling game and Halving game.

Success criterion: *Derive and recall doubles of all numbers to 20*

- Ask the children to look at RCM 16: Doubling game.
- Explain the rules of the game to the children.
  - Each child places their cube on START.
  - Children take turns to roll the die and move their cube the corresponding number of spaces.
  - The child whose turn it is must then double the number their cube lands on.
  - If the other child(ren) in the pair/group decides that the answer is correct, the child collects 1 counter if they are on a white square or 2 counters if they are on a grey square.
  - If the other child(ren) in the pair/group decides that the answer is incorrect, the child collects no counters.
  - The game continues until a child reaches or passes FINISH.
  - Each child then counts all their counters. The winner is the child with most counters.
- As the children play the game, assess whether or not each child can derive and recall doubles for all numbers to 20.

Success criterion: *Derive and recall halves of all even numbers to 40*

- Ask the children to turn the sheet over and look at RCM 17: Halving game.
- Explain the rules of the game to the children.
- Each child places their cube on START.
  - Children take turns to roll the die and move their cube the corresponding number of spaces.
  - The child whose turn it is must then halve the number their cube lands on.
  - If the other child(ren) in the pair/group decides that the answer is correct, the child collects 1 counter if they are on a white square or 2 counters if they are on a grey square.

- If the other child(ren) in the pair/group decides that the answer is incorrect, the child collects no counters.
- The game continues until a child reaches or passes FINISH.
- Each child then counts all their counters. The winner is the child with most counters.

- As the children play the game, assess whether or not each child can derive and recall halves for even numbers to 40.

Success criterion: *Understand that halving is the inverse of doubling*

- Conclude by asking individual children questions similar to the following:
  - **If double 9 is 18, then what is half of 18?**
  - **If half of 20 is 10, what is double 10?**
  - **I'm thinking if a number. I've halved it and the answer is 15. What number was I thinking of? How do you know?**
  - **I'm thinking if a number. I've doubled it and the answer is 16. What number was I thinking of? How do you know?**
  - **What is half of 22? How can you check to make sure you are correct?**
- Continue until each child has sufficiently demonstrated they understand that halving as the inverse of doubling.

AfL

- What is double/twice 16? What is 17 times / multiplied by 2?
- What is half of 14? What is 28 divided by / shared between 2?
- Why are you sure that half of 16 is 8? How can you check if your answer is correct?

What to do for those children who achieve *above* expectation

- Ask children quick-fire questions involving:
  - doubles of all whole numbers to 100
  - doubles of multiples of 5 to 100
  - and the corresponding halves.

What to do for those children who achieve *below* expectation

- Point to numbers on the white squares on RCM 16 and RCM 17 and ask the children to derive and recall doubles of all numbers to 10, and the corresponding halves.

Task 10

Knowing and using number facts

Objective **NC AT 2** **NC Level 2**

- Derive and recall multiplication facts for the 2, 5 and 10 times tables and the related division facts; recognise multiples of 2, 5 and 10

Resources

- RCM 18: 2, 5 and 10 multiplication facts (per child)
- RCM 18: 2, 5 and 10 multiplication facts (enlarged to A3)
- coloured pencil (per child)
- 1–10 die or 0–9 die (per group or pair)
- RCM 20: Multiples of 2, 5 and 10 (enlarged to A3)
- RCM 19: 3, 4 and 6 multiplication facts (for children achieving *above* expectation)
- appropriate apparatus such as 1–100 number square or multiplication square (for children achieving *below* expectation)

Task

- This task involves children playing a game. Prior to the task, decide whether you want the children to play the game as one large group, two smaller groups or in pairs.

Success criterion: *Derive and recall multiplication facts for the 2, 5 and 10 times tables*

- Provide each child with a copy of RCM 18 and a coloured pencil, and each group with either a 1–10 or 0–9 die.
- Explain to the children that they are going to play a game: the Times Tables game.
- If you are using a 0–9 die explain to the children that if they roll a zero on the die that it represents ten not zero.
- Explain the rules of the game to the children:
  - Children take turns to roll the die, e.g. 6.
  - The child whose turn it is chooses whether to multiply the number rolled by 2, 5 or 10, e.g. 5.
  - The child says the calculation, i.e. 6 times 5 equals thirty.
  - The child then draws a cross through the answer, i.e. 30, on the corresponding table.
  - If a child rolls a number they have already rolled three times and therefore cannot cross out another number on their RCM, they miss a turn.
  - The winner is the first child to cross out all the numbers on their RCM or the child who has crossed out the most numbers on their RCM when you announce the end of the game.

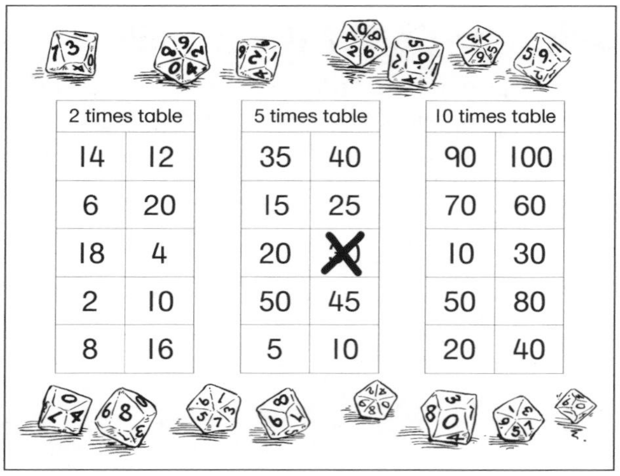

2 times table	
14	12
6	20
18	4
2	10
8	16

5 times table	
35	40
15	25
20	X
50	45
5	10

10 times table	
90	100
70	60
10	30
50	80
20	40

- As the children play the game, assess each child's knowledge of the 2, 5 and 10 multiplication tables. If a child has not won the game by the time you have completed your assessment, tell them the game is over and announce the winner.

Success criterion: *Derive and recall division facts related to the 2, 5 and 10 multiplication facts*

- Place the enlarged copy of RCM 18 on the table in front of the children.
- Say a child's name and point to one of the numbers on the sheet, e.g. 35 on the 5 times table grid, and ask: **Harriet, how many fives are there in 35?**
- Repeat the above for other children in the group, asking questions such as: **Leroy, divide 18 by 2. Yushfa, divide 10 into 70. Harriet, share 8 between 2.**
- Continue until each child has sufficiently demonstrated their ability to derive and recall division facts related to the 2, 5 and 10 multiplication facts.

Success criterion: *Recognise multiples of 2, 5 and 10 up to the tenth multiple*

- Place RCM 20 in the middle of the table for all the children to see. Say: **All these numbers are multiples of 2, 5 or 10. I'm going to point to one of these numbers and I want you to tell me what it is a multiple of.**
- Pointing to one of the numbers, e.g. 45, ask: **Leroy, 45 is a multiple of what number?** (5). **Is it a multiple of another number?**
- Occasionally ask questions such as: **How do you know that 45 is a multiple of 5?**
- Repeat this several times, pointing to different numbers on the sheet.
- Finally, ask: **Harriet, tell me a multiple of 5. Sharon, tell me a multiple of 2**.
- Continue until each child has sufficiently demonstrated their ability to recognise multiples of 2, 5 and 10 up to the tenth multiple.

AfL

- What is 5 times / multiplied by / lots of / groups of / sets of 6?
- What number multiplied by 5 equals 30?
- 5 multiplied by what other number equals 30?
- Tell me two numbers that, when multiplied together, equal 30?
- What is 35 divided by / shared between 5?
- How many twos are there in 12? How do you know?
- Is 15 a multiple of 2? What about a multiple of 5? Is it a multiple of 10? How do you know?
- Tell me another number that is a multiple of 2 …5 …10.

What to do for those children who achieve *above* expectation

- Assess children's understanding of the 3, 4 and 6 times tables and the related division facts using RCM 19 (Level 3).
- Using RCM 20, ask children to identify multiples of 2 and 5 that are beyond the tenth multiple. Ask: **Can you tell me another number that is a multiple of 2? …5? …10? How do you know it is a multiple of 2? …5? …10?**

What to do for those children who achieve *below* expectation

- Allow the children to use appropriate apparatus to work out the answers, e.g. 1–100 number square or a multiplication square.
- Play the game using just one of the tables on RCM 18 to concentrate on one particular multiplication table, i.e. 2, 5 or 10.

Task 11

Knowing and using number facts

Objective **NC AT 2** **NC Level 2**

- Use knowledge of number facts and operations to estimate and check answers to calculations

Resources

- RCM 21: Estimate, calculate and check (per child)
- pencil (per child)
- appropriate apparatus such as 1–100 number square or multiplication square (for children achieving *below* expectation)

Task

NOTE: Prior to the activity, write from four to six different calculations in the first column of RCM 21. The calculations should include addition, subtraction, multiplication and division, and be appropriate to individual children's ability:
– add a one-digit number to a two-digit number
– add a pair of two-digit numbers
– add three or more two-digit numbers
– subtract a one-digit number from a two-digit number
– subtract a pair of two-digit numbers
– multiply a 'teen' number by a one-digit number
– double a two-digit number
– divide a two-digit number by 2, 3, 4, 5 or 10 (including remainders)
– halve a two-digit even number.

Success criteria: *Use knowledge of number facts and operations to estimate answers to calculations*
Use knowledge of number facts and operations to check answers to calculations

- Ask the children to look at the calculations on their sheet.
- Tell the children that for each calculation you want them to:
  – estimate the answer first, writing their estimate and any working out in the second column
  – work out the answer, showing all their working in the third column
  – check their answer, writing any working out in the fourth column.
- As the children work through the calculations on the sheet, ask them questions similar to those below, assessing their ability to use their knowledge of number facts and operations to estimate and check results.

AfL

- What is the approximate answer to this calculation? How did you make your estimate?
- What is the answer to this calculation? How did you work it out?
- How close is the actual answer to your estimate? So do you think that your answer is right? Why?
- Is your answer correct? How can you be so sure?

What to do for those children who achieve *above* expectation

- Ask the children to estimate, calculate and check the following types of calculations:
  – add a pair of two-digit numbers
  – add a two-digit number to a three-digit number
  – add three or more two-digit numbers
  – subtract a pair of two-digit numbers
  – subtract a two-digit number from a three-digit number
  – multiply a two-digit number by a one-digit number
  – double a two-digit number
  – divide a two-digit number by a one-digit number
  – halve a two-digit even number.

What to do for those children who achieve *below* expectation

- Ask the children to estimate, calculate and check the following types of calculations:
  – add a one-digit number to a two-digit number
  – subtract a one-digit number from a two-digit number
  – multiply a pair of one-digit numbers
  – divide a two-digit number by 2, 3, 4, 5 or 10
- Allow the children to use appropriate apparatus to work out the answers, e.g. a 1–100 number square or a multiplication square. Be sure to ask them questions that will help you identify which strategies the children are using to estimate and check.

Task 12

Calculating

Objective **NC AT 2** **NC Level 2**

- **Add or subtract mentally a one-digit number or a multiple of 10 to or from any two-digit number; use practical and informal written methods to add and subtract two-digit numbers**

Resources

- RCM 4: 1–20 number cards
- RCM 5: Two-digit number cards
- RCM 15: Multiples of 10 number cards
- pencil and paper (per child)
- selection of different counting apparatus such as 1–100 number square, counters, interlocking cubes, Base10 material
- RCM 6: Three-digit number cards (for children achieving *above* expectation)

Task

NOTES:

– This task is best undertaken with only two or three children.

– This task involves a number of Success criteria. It is advisable to choose just one or two at a time.

– Have pencils and paper available for the children to use. However, do not draw attention to these. If appropriate, allow the children to use the pencil and paper to carry out informal written methods to add and subtract two-digit numbers, otherwise encourage the children to use mental methods to work out the answers to the calculations.

– Similarly, have to hand the selection of different counting apparatus. Allow the children to use these if necessary.

- Prior to the task, shuffle and place each of the following cards into separate piles:
  - 1–9 number cards from RCM 4
  - 10–20 number cards from RCM 4
  - two-digit number cards from RCM 5
  - multiples of 10 number cards from RCM 15.
- Place on the table the pencils and paper, and the selection of different counting apparatus.

Success criterion: *Add mentally a one-digit number to any two-digit number*

- Using the piles of 1–9 number cards, 10–20 number cards and two-digit number cards, place a one-digit number card, e.g. 6, and a two-digit card, e.g. 67, in front of each child.
- Say: **Marsha, add these two numbers together.**
- Repeat the above several times using appropriate mathematical vocabulary, e.g. **What is the sum of these two numbers? What is the total of the numbers on these two cards? What is this number plus this number? What is 67 and 6 more?**
- Occasionally ask: **How did you work that out? How did you get that answer?**
- Continue until each child has sufficiently demonstrated their ability to add mentally a one-digit number to any two-digit number.

Success criterion: *Add mentally a multiple of 10 to any two-digit number*

- Using the piles of multiples of 10 number cards, 10–20 number cards and two-digit number cards, place a multiple of 10 number card, e.g. 40, and a two-digit card, e.g. 57, in front of each child.
- Ask: **Gabrielle, what is the sum of these two numbers?**
- Repeat the above several times using appropriate mathematical vocabulary, e.g. **Add these two numbers together. What is the total of the numbers on these two cards? What is this number plus this number? What is 40 more than 57?**
- Occasionally ask: **How did you work that out? How did you get that answer?**
- Continue until each child has sufficiently demonstrated their ability to add mentally a multiple of 10 to any two-digit number.

Success criterion: *Use practical and informal written methods to add two-digit numbers*

- Using the piles of 10–20 number cards and two-digit number cards, place two two-digit number cards in front of each child, e.g. 26 and 52.
- Say: **Tina, add these two numbers together for me.**
- Repeat the above several times using appropriate mathematical vocabulary, e.g. **What is the sum of the two numbers on your cards? What is the total of these two numbers? What is this number plus this number? What is 26 and 52 more?**
- Occasionally ask: **How did you work that out? How did you get that answer? Show me your working out. Can you explain it to me?**
- Continue until each child has sufficiently demonstrated their ability to use practical and informal written methods to add two-digit numbers.

Success criterion: *Subtract mentally a one-digit number from any two-digit number*

- Using the piles of 1–9 number cards, 10–20 number cards and two-digit number cards, place a one-digit number card, e.g. 8, and a two-digit card, e.g. 35, in front of each child.
- Ask: **Tina, what is 35 take away 8?**
- Repeat the above several times using appropriate mathematical vocabulary, e.g. **What is the difference between these two numbers? What is 35 minus 8? What is 8 less than 35? Subtract 8 from 35.**
- Occasionally ask: **How did you work that out? How did you get that answer?**
- Continue until each child has sufficiently demonstrated their ability to subtract mentally a one-digit number from any two-digit number.

Success criterion: *Subtract mentally a multiple of 10 from any two-digit number*

- Using the piles of multiples of 10 number cards, 10–20 number cards and two-digit number cards, place a multiple of 10 number card, e.g. 30, and a two-digit number card, e.g. 67, in front of each child.
- Ask: **Gabrielle, what is 67 subtract 30?**
- Repeat the above several times using appropriate mathematical vocabulary, e.g. **What is 67 take away 30? What is the difference between these two numbers? What is 67 minus 30? What is 30 less than 67? Subtract 30 from 67.**
- Occasionally ask: **How did you work that out? How did you get that answer?**
- Continue until each child has sufficiently demonstrated their ability to subtract mentally a multiple of 10 from any two-digit number.

Success criterion: *Use practical and informal written methods to subtract two-digit numbers*

- Using the piles of 10–20 number cards and two-digit number cards, place two two-digit number cards in front of each child, e.g. 41 and 85.
- Ask: **Marsha, what is the difference between 41 and 85?**
- Repeat the above several times using appropriate mathematical vocabulary, e.g. **Subtract 41 from 85. What is 85 minus 41? What number is 41 less than 85? Take the smaller number away from the larger number. Take 41 away from 85.**
- Occasionally ask: **How did you work that out? How did you get that answer? Show me your working out. Can you explain it to me?**
- Continue until each child has sufficiently demonstrated their ability to use practical and informal written methods to subtract two-digit numbers.

AfL

- What is 27 add 7? What number facts might you use to help you work this out? What do you need to add to 27 to get to the next multiple of 10? How might you partition 27 to help you?
- What is 46 plus 40? How did you work this out?
- What is 54 add 38? How did you get that answer?
- Show me how you could work out the answer to 38 minus 7? …56 minus 30? What about 84 minus 56?
- Can you work out your answer in a different way? Which way do you find most helpful? Why?

What to do for those children who achieve *above* expectation

- Using the three-digit number cards from RCM 6 and the two-digit number cards from RCM 5, ask the children to use informal and written methods to add and subtract two-digit and three-digit numbers, e.g. 522 ± 38, 258 ± 169 (Level 3).

What to do for those children who achieve *below* expectation

- Encourage the children to use practical methods using the counting apparatus and informal jottings.

Task 13

Calculating

Objective **NC AT 2** **NC Level 2**

- Understand that subtraction is the inverse of addition and vice versa and use this to derive and record related addition and subtraction number sentences

Resources

- RCM 22: Addition and subtraction facts (enlarged to A3)

Task

Success criteria: *Understand that subtraction is the inverse of addition and vice versa*
Use this to derive and record related addition and subtraction number sentences

- Show the children RCM 22.
- Explain to the children that the sheet shows different addition and subtraction calculations.
- Pointing to one of the addition facts, e.g. 3 + 6 = 9, ask a child to say a related subtraction fact, i.e. 9 − 3 = 6 or 9 − 6 = 3. Point and say: **Yolanda, look at this addition number fact, 3 add 6 equals 9. Tell me a subtraction fact that uses the same three numbers.**
- Pointing to one of the subtraction facts, e.g. 10 − 3 = 7, ask another child to say a related addition fact, i.e. 3 + 7 = 10 or 7 + 3 = 10. Point and say: **William, look at this subtraction number fact, 10 minus 3 equals 7. I want you to tell me a related addition fact.**
- Repeat the above until each child has sufficiently demonstrated their ability to state a subtraction fact corresponding to a given addition fact and an addition fact corresponding to a given subtraction fact.

AfL

- What addition facts can you use to help you work out the answer to 16 minus 5?
- 14 add 5 equals 19. Tell me three more calculations using these three numbers.
- How do you know, without calculating, that they are correct?
- I think of a number and add 5. The answer is 12. What is my number?
- 17 − 8, 13 − 6. What addition facts can you use to help you calculate these? Explain how the addition facts help you.
- Using the numbers 3, 6 and 9 tell me an addition number sentence. Can you tell me another addition number sentence? Can you tell me a related subtraction number sentence? Can you tell me another subtraction number sentence?

What to do for those children who achieve *above* expectation

- Pointing to an addition fact, e.g. $5 + 2 = 7$, ask the children to state the two related subtraction facts, i.e. $7 - 5 = 2$ and $7 - 2 = 5$.
- Pointing to a subtraction fact, e.g. $6 - 4 = 2$, ask the children to state the two related addition facts, i.e. $4 + 2 = 6$ and $2 + 4 = 6$.
- Pointing to the flags in the bottom three rows of the RCM, ask the children to state the subtraction fact(s) corresponding to a given addition fact and the addition fact(s) corresponding to a given subtraction fact.

What to do for those children who achieve *below* expectation

- Pointing to the flags in the top four rows of the RCM, ask the children to state a subtraction fact corresponding to a given addition fact and an addition fact corresponding to a given subtraction fact for addition and subtraction number facts to 10.

Task 14

Calculating

Objective **NC AT 2** **NC Level 2**

- Represent repeated addition and arrays as multiplication, and sharing and repeated subtraction (grouping) as division; use practical and informal written methods and related vocabulary to support multiplication and division, including calculations with remainders

Resources

- RCM 12: Array cards
- RCM 4: 1–20 number cards
- RCM 5: Two-digit number cards
- pencil and paper (per child)
- about 50 counters
- large sheets of paper and a marker
- selection of different counting apparatus such as multiplication square, counters, interlocking cubes

Task

NOTES:

– This task is best undertaken with only two or three children.

– This task involves a number of Success criteria. It is advisable to choose just one or two at a time.

- Prior to the task:
  - shuffle the array cards from RCM 12 and the number cards from RCM 4 and RCM 5 and place each set of cards in a pile of their own
  - provide each child with a pencil and a piece of paper.

Success criterion: *Represent repeated addition and arrays as multiplication*

- Place one of the array cards from RCM 12 in front of each child, e.g.

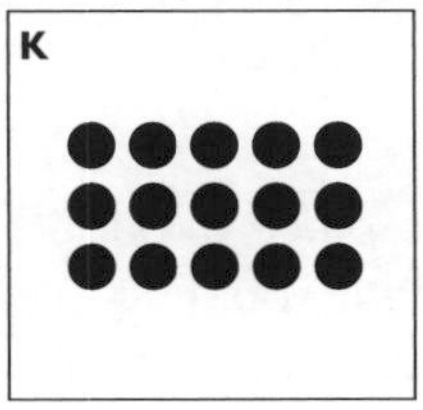

- Ask: **Jane, how many dots are there? How can you work it out without counting them all? I want you to write down a number sentence to record your method using the multiplication or addition symbol.**
- Depending on which operation the child uses, i.e. multiplication or addition, ask them to write another number sentence using the other operation.
- Say: **Good, 5 add 5 add 5 equals 15.** Ask: **Now, can you write down another number sentence using the multiplication symbol?**

  NOTE: Accept any of the following calculations: $5 + 5 + 5 = 15$ and $3 + 3 + 3 + 3 + 3 = 15$; $3 \times 5 = 15$ and $5 \times 3 = 15$.
- Repeat the above until each child has sufficiently demonstrated that they can represent repeated addition and arrays as multiplication.

Success criterion: *Represent sharing as division*

- Draw three circles on one of the sheets of A3 paper and place 12 counters near the circles, e.g.

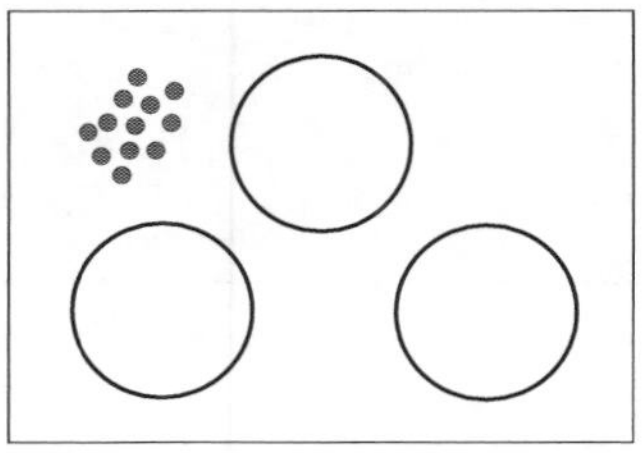

- Referring to one child in the group, ask: **James, how many counters are there?** Say: **James, I want you to share these counters equally between three groups**.
- When the child has done the task, ask: **James, how many counters are there in each group? Can you tell me this as a division calculation?**
- Repeat the task for the same child, drawing a different number of circles and using a different number of counters.
- Repeat the above for each child until they have sufficiently demonstrated their ability to represent sharing as division.

Success criterion: *Represent repeated subtraction (grouping) as division*

- Place 15 counters on the table.
- Referring to one child in the group, ask: **Jane, how many counters are there?** Say: **Jane, I want you to arrange these counters into groups of five.**
- When the child has done the task, ask: **Jane, how many groups did you make? Can you tell me this as a division calculation?**
- Repeat the task for the same child using a different number of counters and asking them to arrange them into groups of a different size.
- Repeat the above for each child until they have sufficiently demonstrated their ability to represent repeated subtraction (grouping) as division.

Success criterion: *Use practical and informal written methods and related vocabulary to support multiplication*

- Prior to this part of the task:
  - ensure that each child has a pencil and a piece of paper
  - place on the table the selection of different counting apparatus
  - decide which two-digit number card to give each child, and what one-digit number to ask them to multiply the two-digit number by:

	Easy	**Difficult**
Number cards	'teen' number card from RCM 4	two-digit number card from RCM 5
One-digit number	2, 3, 4 or 5	6, 7, 8 or 9

- Place a 'teen' number card from RCM 4 or a two-digit number card from RCM 5 in front of each child.
- Say: **I'm going to tell each of you a number. I then want you to multiply the number on your card by the number I tell you. You can use some of the counting materials or use your sheet of paper to make any jottings.**

- Once the children have done this, ask: **What is the answer to your calculation? How did you get that answer?**
- Continue until each child has sufficiently demonstrated their ability to use practical and informal written methods to support multiplication.

Success criterion: *Use practical and informal written methods and related vocabulary to support division, including calculations with remainders*

- Prior to this part of the task:
  - ensure that each child has a pencil and a piece of paper
  - place on the table the selection of different counting apparatus
  - decide which two-digit number card to give each child, and what one-digit number to ask them to divide the two-digit number by:

	Easy	Difficult
Number cards	'teen' number card from RCM 4	two-digit number card from RCM 5
One-digit number	2, 3, 4 or 5	6, 7, 8 or 9
	Use numbers that do not result in division calculations involving remainders	Use numbers that will result in division calculations involving remainders

- Place a 'teen' number card from RCM 4 or a two-digit number card from RCM 5 in front of each child.
- Say: **I'm going to tell each of you another number, but this time I want you to divide the number I tell you into the number on your card. You can use any of the counting materials or use your sheet of paper to write down any jottings.**
- Once the children have done this, ask: **What is the answer to your calculation? How did you work it out?**
- Continue until each child has sufficiently demonstrated their ability to use practical and informal written methods to support division, including calculations with remainders.

AfL

- How many dots are there? How can you work it out without counting them all?
- What number sentence can you write to record your method using the multiplication/addition symbol?
- Can you write down another number sentence using the addition/multiplication symbol?
- Is there another number sentence you could make?
- How would you arrange these 20 counters into equal rows? How many rows are there? How many counters are there in each row? Write down a number sentence to show how you have arranged the counters.
- What is the answer to 12 divided by 3? What about 14 divided by 3?
- Tell me some numbers that divide exactly by 2. …5. …10. How do you know?
- If 7 times 4 is 28, what is 28 divided by 7? What other facts do you know?
- If I multiply a number by 5 and then divide the number by 5, what happens?
- If I divide a number by 5 and then multiply the number by 5, what happens?

What to do for those children who achieve *above* expectation

- Without using any counters, ask the children to describe a way they could make equal rows of 18, 20, 24, 36… counters. Ask: **How would you record that? Is there another way you could record it? Which do you think is the best method of recording? Why? Can you think of another way you could arrange the counters? How would you record it?**
- Ask the children to identify the corresponding division fact for a multiplication fact in the 2, 3, 4, 5 and 6 times tables, e.g. Say: **Steven, 7 times 6 is 42.** Ask: **What is a related division fact for this calculation?** Say: **Tell me a multiplication fact you know and a related division fact.**
- When asking the children to use practical and informal written methods to support multiplication and division, including calculations with remainders, use the grids on pages 44 and 45. Also encourage the children to use informal written methods.

What to do for those children who achieve *below* expectation

- Only use those array cards that describe a multiplication fact for the 2, 5 or 10 times tables.
- Only ask questions involving division as sharing and grouping that are related to the 2, 5 and 10 times tables.
- Using the large sheets of paper and the counters, ask the children to show you a multiplication fact, e.g. 3×2. Ask: **So, if 3 groups of 2 is 6, what is 6 divided into 3 groups?** Repeat for other multiplication and related division facts in the 2, 5 and 10 times tables.
- When asking the children to use practical and informal written methods to support multiplication and division, for calculations with no remainders, use the grids on pages 44 and 45. Also encourage the children to use practical methods using the counting apparatus.

Task 15

Calculating

Objective **NC AT 2** **NC Level 2**

- **Use the symbols +, −, ×, ÷ and = to record and interpret number sentences involving all four operations; calculate the value of an unknown in a number sentence, e.g. □ ÷ 2 = 6, 30 − □ = 24**

Resources

- RCM 4: 1–20 number cards
- RCM 5: Two-digit number cards
- RCM 9: Symbol cards
- RCM 15: Multiples of 10 number cards
- RCM 23: Operator cards

Task

- Prior to the task, place each of the following cards in separate piles:
  - 1–9 number cards from RCM 4
  - 10–20 number cards from RCM 4
  - two-digit number cards from RCM 5
  - multiples of 10 number cards from RCM 15
  - equals (=) symbol cards from RCM 9
  - addition (+) symbol cards from RCM 23
  - subtraction (−) symbol cards from RCM 23
  - multiplication (×) symbol cards from RCM 23
  - division (÷) symbol cards from RCM 23
  - square shape cards from RCM 9 and circle and triangle shape cards from RCM 23.

  You will not need the greater than (>) and less than (<) symbol cards from RCM 9 for this task.

Success criterion: *Use the symbols +, −, ×, ÷ and = to record and interpret number sentences involving all four operations*

- Place the piles of number cards from RCMs 4, 5 and 15, and the addition, subtraction, multiplication, division and equals symbol cards from RCM 9 and RCM 23 face up in the middle of the table.
- Ask a calculation of one of the children, e.g. ask: **Louise, what is 4 add 7?** Then ask the same child to use the number cards and symbol cards to make this calculation. Say: **Louise, I want you to use these number cards and symbol cards to show me this as a number sentence.** i.e. [4] [+] [7] [=] [11]
- Occasionally use the cards to make a calculation of your own, e.g. [3] [×] [4] [=] [12] and ask the children to interpret the number sentence. Ask: **Simone, can you read me this number sentence / calculation?**
- Continue asking a variety of addition, subtraction, multiplication and division calculations until each child has sufficiently demonstrated their ability to use the symbols +, −, ×, ÷ and = to record and interpret number sentences involving all four operations.

Success criterion: *Calculate the value of an unknown in a number sentence*

- Place the pile of square, circle and triangle shape cards from RCM 9 and RCM 23 alongside the other cards on the table.
- Use the cards to create a calculation where the children are required to calculate the value of an unknown in a number sentence, e.g. [6] [+] [□] [=] [15], [○] [−] [4] [=] [9], [△] [=] [20] [+] [30], [6] [=] [12] [÷] [○], [9] [□] [6] [=] [15], [5] [○] [4] [=] [20].
- Ask: **Simone, can you tell me the value of the triangle? What should replace the square to complete this number sentence? Which card do you need to replace the circle card with to make this number sentence true?**
- Repeat the above, creating a variety of addition, subtraction, multiplication and division calculations with the unknown value in different positions in the number sentence.
- Continue until each child has sufficiently demonstrated their ability to calculate the value of an unknown in a number sentence.

AfL

- What number goes in the box to make this calculation correct □ ÷ 2 = 5? How do you know?
- Can you make three different number sentences using 12, 5 and 17 with = and any of the four operation symbols?
- Can you change the three numbers to make this into a different problem for someone else to solve? How will you know if their answer is correct?
- 16 + □ = 25. What is the missing number? How do you know? What subtraction could you do to find the answer?
- How many different ways can you find of adding three numbers to make 12?
- Explain how you worked out the missing number in this number sentence: 35 ÷ □ = 7
- How can you record the solution to this problem?
- I am thinking of a number. I divide it by 5 and the answer is 3. What is my number?
- Make up some 'missing-number' problems for others to solve.

What to do for those children who achieve *above* expectation

- Ask the children to record and interpret number sentences involving:
  - addition and subtraction number facts to 20
  - addition and subtraction of a one-digit number to or from any two-digit number
  - addition and subtraction of pairs of two-digit numbers
  - sums and differences of multiples of 10
  - 2, 3, 4, 5, 6 and 10 times tables and the related division facts.

What to do for those children who achieve *below* expectation

- Only ask the children to record and interpret number sentences involving:
  - addition and subtraction number facts to 10
  - 2, 5 and 10 times tables and the related division facts.

Task 16

Understanding shape

Objective **NC AT 3** **NC Level 2**

- **Visualise common 2-D shapes and 3-D solids; identify shapes from pictures of them in different positions and orientations; sort, make and describe shapes, referring to their properties**

Resources

- RCM 24: Shapes and solids (enlarged to A3)
- RCM 25: Shape property cards
- set of regular and irregular polygons
- set of geometric solid shapes
- ruler (per child)
- pencil (per child)
- square dot paper (per child)
- triangular dot paper (per child)
- set of polyhedrons

Task

NOTES:

– This task is best undertaken with only one or two children.

– This task involves a number of Success criteria. It is advisable to choose just one or two at a time.

Success criterion: *Visualise common 2-D shapes and 3-D solids*

- Describe a shape to the children, e.g. say: **I'm thinking of a shape. It has four sides and four right angles. The opposite sides of this shape are the same length.** Ask: **What shape am I thinking of?**
- Repeat, describing other 2-D shapes and 3-D solids to the children.
- Continue until each child has sufficiently demonstrated their ability to visualise the following common 2-D shapes and 3-D solids: circle, triangle, square, rectangle, pentagon, hexagon, octagon, cube, cuboid, pyramid, cone, cylinder and sphere.

Success criterion: *Identify shapes from pictures of them in different positions and orientations*

- Show the children RCM 24.
- Ask each child in turn to point to a particular shape. Ask: **Lisa, can you point to a hexagon? Louise, can you point to another hexagon? Sam, point to a cuboid.**
- Also ask each child to name certain shapes and solids, e.g. point to a particular shape or solid and ask: **What is this shape called? What is the name of this shape?**
- Repeat for other 2-D shapes and 3-D solids.
- Continue until each child has sufficiently demonstrated their ability to identify the following 2-D shapes and 3-D solids: circle, triangle, square, rectangle, pentagon, hexagon, octagon, cube, cuboid, pyramid, cone, cylinder and sphere.

Success criterion: *Describe 2-D shapes and 3-D solids referring to their properties*

- With RCM 24 still on the table, place one of the Shape property cards from RCM 25 face up on the table, e.g. Has 4 sides.
- Ask a child to name a shape on the table that has four sides. Ask: **Lisa, which of these shapes has four sides? Are there any others? Any more?**
- Repeat the above, placing the Shape property cards one at a time on the table and asking individual children to name the shape(s) that has/have that property.
- Repeat the above, placing two or more Shape property cards on the table and asking individual children to name the shapes that have both those properties, e.g. a) Has 4 sides, Sides are straight, b) Has a square face, Has a triangular face.
- Next, point to a shape or solid on RCM 24 and ask children to describe its properties. Ask questions such as: **Lee, what is this shape called? Can you describe it to me?**
- If children have difficultly in describing the properties of a shape/solid, ask questions such as: **How many sides / angles / lines of symmetry does this 2-D shape have? How many faces / edges / corners does this 3-D solid have?**
- Continue until each child has sufficiently demonstrated their ability to describe a range of 2-D shapes and 3-D solids.

Success criterion: *Sort 2-D shapes and 3-D solids referring to their properties*

- Place the set of regular and irregular polygons and geometric solid shapes on the table.
- Ask individual children to sort the shapes using the following criteria:
  - 2-D shapes and 3-D solids
  - the number of corners
  - the number of sides
  - whether the sides are straight or curved
  - the number of faces
  - the shape of the face
  - whether the faces are flat or curved
- Say: **Lee, sort these shapes into 2-D shapes and 3-D solids. Lisa, sort these shapes into those that have less than 5 corners and 5 or more corners.**
- Conclude by asking individual children to sort the shapes using their own criteria. Ask: **Simon, how would you sort these shapes? Why did you sort them in this way? How else could you sort them? Could you sort them in another way?**
- Continue until each child has sufficiently demonstrated their ability to sort 2-D shapes and 3-D solids referring to their properties.

Success criterion: *Draw 2-D shapes referring to their properties*

- Provide each child with a ruler, pencil, square dot paper and triangular dot paper.
- Ask the children to draw a 2-D shape. Say: **Jake, I want you to draw a rectangle for me.**
- You may wish to ask the children simply to draw a particular polygon or be more specific and ask them to draw a particular 'regular' polygon.
- Repeat, asking the children to draw another polygon.
- Continue until each child has sufficiently demonstrated their ability to draw 2-D shapes.

Success criterion: *Make 3-D solids referring to their properties*

- Place the set of polyhedrons in the middle of the table.
- Ask the children to make a 3-D solid. Say: **Lottie, I want you to use these polyhedrons and make a cube for me.**
- Repeat, asking the children to make another 3-D solid.
- Continue until each child has sufficiently demonstrated their ability to make 3-D solids.

AfL

- What is this shape/solid called?
- Point to a cuboid. …a regular pentagon.
- How many right angles does a regular pentagon/heptagon… have?
- How many lines of symmetry does a rectangle/octagon… have?
- How many faces does this shape have?
- Describe a pentagon to me.
- Is this shape a rectangle? How do you know? Tell me something that is rectangular.
- Which shape has 8 corners and 6 square faces?
- Tell me a shape that has one curved edge.

What to do for those children who achieve *above* expectation

- Point to particular shapes on RCM 24 and ask the children to tell you all they know about the shape. If necessary, ask prompting questions such as: **What is this shape called? Is it regular or irregular? How many angles does it have? Point to them. How many of these angles are right angles? Point to them. How many lines of symmetry does this shape have? Show me.**
- When asking the children to describe and sort the shapes, encourage them to talk about the following:
  - whether or not the shape has reflective symmetry
  - the number of sides/edges and corners (vertices)
  - the number of parallel and perpendicular sides
  - whether sides/edges are the same length
  - whether or not angles are right angles
  - the number and shapes of faces.

What to do for those children who achieve *below* expectation

- Only ask the children to name and describe the properties of the following shapes:
  - 2-D shapes: square, rectangle, triangle, circle
  - 3-D solids: cube, cuboid, sphere, cylinder, cone.
- If appropriate, substitute RCM 24 for the set of regular and irregular polygons and the set of geometric solid shapes.

Task 17

Understanding shape

Objective **NC AT 3** **NC Level 2**

- Identify reflective symmetry in patterns and 2-D shapes and draw lines of symmetry in shapes

Resources

- selection of different apparatus that can be used to make symmetrical patterns, e.g. counters, interlocking cubes, pinboard, pattern blocks, beads and string
- set of regular polygons
- mirror
- pencil (per child)
- ruler (per child)
- RCM 26: Shape cards
- squared paper (for those children achieving *above* expectation)
- scissors (for those children achieving *below* expectation)

Task

Success criterion: *Identify reflective symmetry in patterns*

- Using a ruler and the counters, make a symmetrical pattern, e.g.

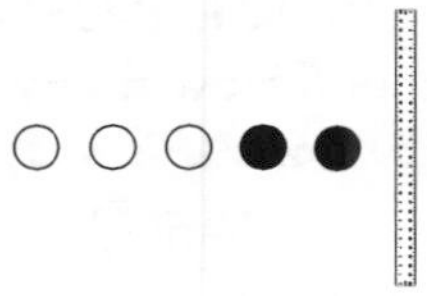

- Ask the children to complete the pattern so that it is symmetrical, i.e.

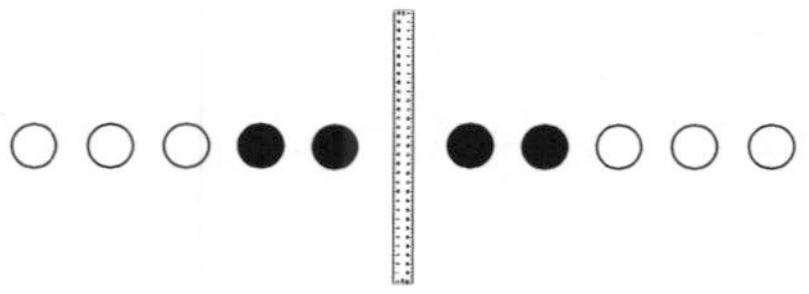

- Repeat for other patterns using two or more colours, e.g.

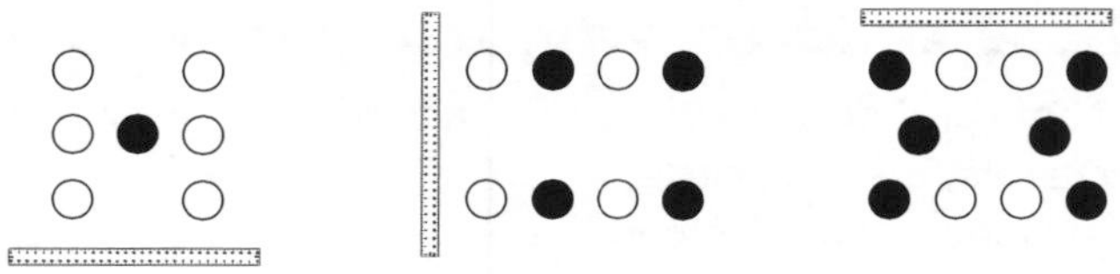

- Repeat using other apparatus, e.g.

Interlocking cubes

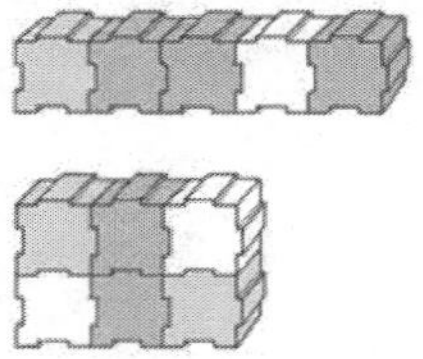

Pinboards

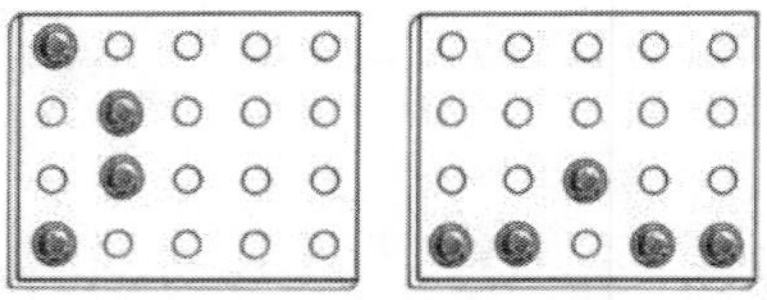

Pattern blocks

Beads

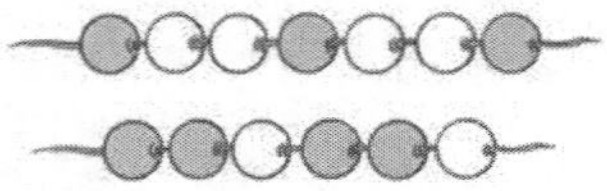

- For those patterns that can be completed in different ways to show different reflective symmetries, e.g. vertical, horizontal or diagonal, say: **Let's look at the original pinboard design that I gave you.** Ask: **Can you make a different pattern that is also symmetrical?**

- Occasionally, use the apparatus yourself to make a pattern that is either symmetrical or asymmetrical and ask: **Is this pattern symmetrical? How can you tell?** If the pattern is asymmetrical, ask: **How can you change it to make it symmetrical?**
- Continue until each child has sufficiently demonstrated their ability to identify reflective symmetry in patterns.

Success criterion: *Identify reflective symmetry in 2-D shapes*

- Place the set of regular polygons and the mirror on the table.
- Pointing to a shape, ask a child to identify a line of symmetry. Ask: **Hayley, can you show me a line of symmetry in this shape? Kevin, can you find another line of symmetry in this square? Rachel, show me a line of symmetry in this rectangle.**
- Continue until each child has sufficiently demonstrated their ability to identify reflective symmetry in 2-D shapes.

Success criterion: *Draw lines of symmetry in shapes*

- Provide each child with a pencil and a ruler and a shape card from RCM 26.
- Ask the children to use their ruler to draw a line of symmetry on their shape. Say: **Using your ruler, I want you to draw for me a line of symmetry on this shape.**
- Once the children have done this, ask: **Does your shape have another line of symmetry? Can you draw it for me? Can you see any other lines of symmetry?**
- Continue until each child has sufficiently demonstrated their ability to draw lines of symmetry in 2-D shapes.

AfL

- Look at this pattern. Can you complete it so that it is symmetrical? Can you complete the pattern in a different way so that it also symmetrical?
- How do you know this shape is symmetrical?
- Point to the line of symmetry.
- Show me a line of symmetry on this shape? Does it have any other lines of symmetry?
- How many lines of symmetry does this shape have?

What to do for those children who achieve *above* expectation

- Draw a simple shape on squared paper and ask the children to draw the reflection in a mirror line along one side (Level 3), e.g.

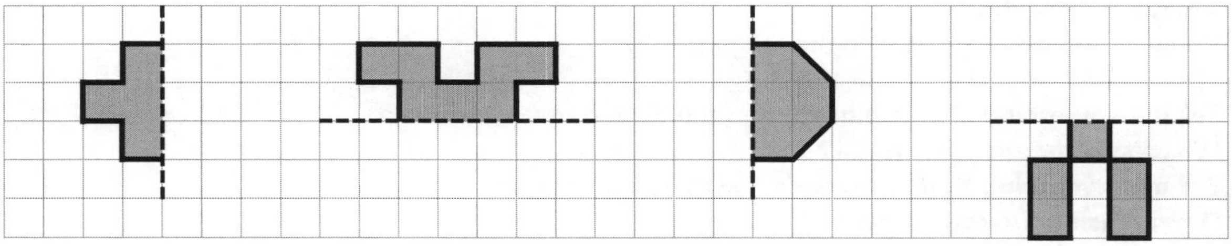

What to do for those children who achieve *below* expectation

- Allow the children to use scissors to cut out the shape on their shape card and then to fold the shape to show the line(s) of symmetry.

Task 18

Understanding shape

Objective **NC AT 3** **NC Level 2**

- Follow and give instructions involving position, direction and movement

Resources

- RCM 27: Position, direction and movement (enlarged to A3)
- coloured pencils
- 1–6 die
- blank die labelled: up, down, left, right, joker, joker
- different button (per child)
- container of counters
- pencil
- 2 geostrips fastened together, i.e.

- another copy of RCM 27: Position, direction and movement (enlarged to A3) (for children achieving *above* expectation)

Task

- Place RCM 27 and the coloured pencils on the table in front of the children. Ensure that all the children are able to see the sheet from the same perspective.
- Briefly discuss with the children any of the illustrations they may not recognise.

Success criterion: *Follow and give instructions involving position*

- Tell the children to ignore for the moment the coins and the 'start' box.
- Pointing to one of the objects on the grid that is not a coin, ask individual children questions similar to the following:
  - **Tell me something that is to the left of the bus.**
  - **Tell me something that is higher than the ...**
  - **What is next to the ...?**
  - **Point to something that is on the edge of the grid.**
  - **What is below the ...?**
  - **Point to something that is higher than the ...**
  - **What is between the ... and the ...?**
  - **Tell me something that is higher than the ... and lower than the ...**
  - **What is to the right of the ...?**
  - **Tell me something that is towards the middle of the sheet.**
  - **Draw a fish below the flower.**
  - **Draw a face to the left of the shoe.**
- Ask children to choose two objects on the grid and describe their position in relation to each other. Ask: **Jerry, point to two objects on the grid. Good. What can you tell me about the position of these two objects?**
- Repeat the above until each child has sufficiently demonstrated that they can follow and give instructions involving position.

Success criterion: *Follow and give instructions involving direction*

- With RCM 27 still on the table, show the children the dice, the different buttons and the container of counters.
- Show the children how the blank die is marked with the following: up, down, left, right, joker, joker.
- Provide each child with a different button.
- Explain to the children that they are going to play a game.
- Explain the rules of the game to the children.
  - Children place their button on 'start'.
  - They take turns to roll the dice and move their button accordingly. For example, if a child rolls 4 and down, then they have to move their button down four spaces.
  - If a child rolls a joker, then you (the adult), will say in which direction to move the button, i.e. either up, down, left or right.
  - If a child lands on a space that shows 1, 2 or 3 coins, then they collect that number of counters.
  - If the position of a child's button is towards the top, bottom, left or right of the grid and, on their next turn, the roll of the dice means they will go off the grid, then they miss that turn.
  - The game continues in this way.
  - The winner is the first child to collect 10 counters or the child with the most counters after a predetermined time.
- As the children play the game, assess their ability to follow and give instructions involving direction.
- Now place all the resources used so far to one side.

Success criterion: *Follow and give instructions involving movement*

- Show the children the pencil and the two geostrips fastened together.
- Ask individual children questions similar to the following:
  - **Rachel, move this pencil clockwise / anticlockwise.**
  - **Michael, roll / slide this pencil along the table.**
  - **Look at the geostrips. Move this geostrip to show a quarter / half / whole turn.**
  - **Rachel, show me a right angle / straight line using these geostrips.**
- Repeat the above until each child has sufficiently demonstrated that they can follow and give instructions involving movement.

AfL

- Describe for me the position of the banana.
- What is above / below / next to / to the left of / to the right of the tree?
- Describe for me your journey to school.
- Tell me some things that move? How do they move?

What to do for those children who achieve *above* expectation

- Using the other copy of RCM 27, write the letters A to M underneath the 13 boxes in the bottom row, and the numbers 1 to 13 beside the first column of boxes, e.g.

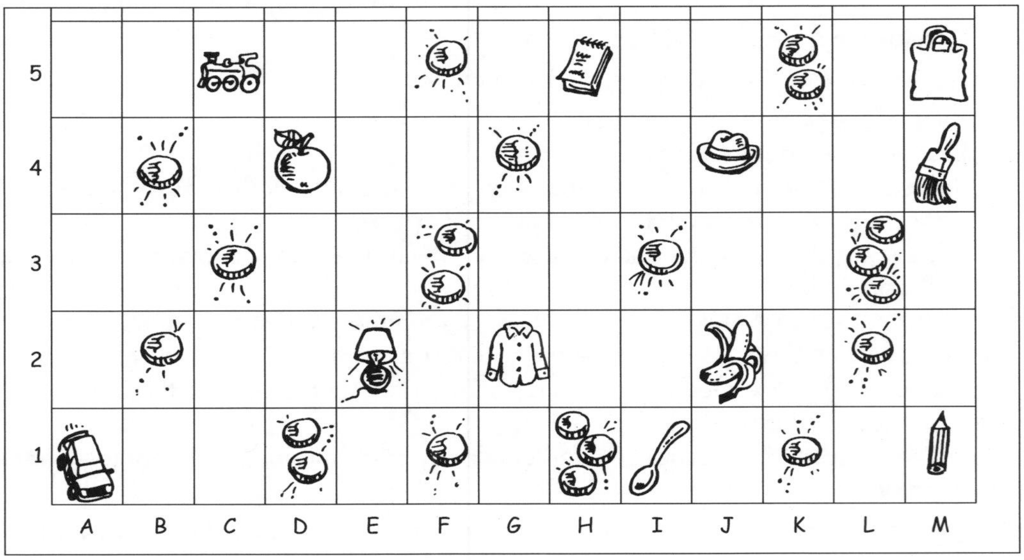

Ask the children to describe the position of certain objects on the grid using the co-ordinates, e.g. ask: **What is in box E8?** (tree) **In which box is the hat?** (J4)

- Ask the children to describe:
  - directions using the four compass directions: N, S, E, W
  - movement, recognising whole, half and quarter turns.

What to do for those children who achieve *below* expectation

- Ask the children to describe:
  - the position of objects in the classroom
  - familiar journeys, e.g. how to walk from the classroom to the hall, their journey to and from school
  - things that move.

Task 19

Understanding shape

Objective **NC AT 3** **NC Level 2**

- Recognise and use whole, half and quarter turns, both clockwise and anticlockwise; know that a right angle represents a quarter turn

Resources

- selection of objects that can be turned, e.g. picture from a magazine, photo, book, shoe, pencil, scissors, paint brush
- RCM 24: Shapes and solids (enlarged to A3)
- 2 geostrips fastened together, i.e. 
- set-square (for children achieving *above* expectation)
- pencil and paper (for children achieving *above* expectation)
- RCM 26: Shape cards (for children achieving *below* expectation)

Task

NOTE: This task is best undertaken with only one or two children.

Success criterion: *Recognise and use whole, half and quarter turns, both clockwise and anticlockwise*

- Place the selection of objects that can be turned on the table. Ensure that all the children are able to see the objects from the same perspective.
- Take one of the objects, e.g. a picture from a magazine, and place it right way up in front of the children.
- Say: **Look at this picture. Close your eyes while I turn it.** Turn the picture a whole, half or quarter turn, clockwise or anticlockwise. Say: **Now open your eyes.** Ask: **What did I do? Are you sure? How could you check?**
- Repeat the above several times using the other objects, turning them a whole, half or quarter turn, clockwise or anticlockwise.
- Next, place one of the objects in front of one of the children and say: **Turn this picture half a turn clockwise. Now turn the picture a quarter turn anticlockwise.** Ask: **How can we get it back to where it started from? Is there any other way?**
- Repeat the above for other children in the group, asking them to turn their object a whole, half or quarter turn, clockwise or anticlockwise.
- Continue until each child has demonstrated their ability to recognise and use whole, half and quarter turns, both clockwise and anticlockwise.

Success criterion: *Know that a right angle represents a quarter turn*

- Show the children the two geostrips fastened together.
- Ask individual children questions similar to the following:
  - **Look at these geostrips. Move this geostrip to show a quarter turn.**
  - **Rachel, show me a right angle using these geostrips.**
  - **What can you tell me about the amount of turn in a right angle?**
  - **What can you tell me about a right angle?**

Success criterion: *Identify right angles in shapes and in the environment*

- Show the children RCM 24.
- Draw children's attention to the 2-D shapes only and briefly name and discuss some of the 2-D shapes on the RCM.
- Ask each child in turn to point to a shape and identify a right angle. Say: **Lena, look at all the shapes on the sheet. Point to a right angle in one of the shapes.**
- Also ask the children to identify right angles in the environment. Say: **Lena, find a right angle in this room.**
- Repeat until each child has sufficiently demonstrated their ability to identify right angles in shapes and in the environment.

AfL

- Turn this picture half a turn anticlockwise. Now turn the picture a quarter turn clockwise. How can we get it back to where it started from? Is there any other way?
- Look at this picture. Close your eyes while I turn it. Now open your eyes. What did I do? Are you sure? How could you check?
- Use these geostrips to show me what a right angle looks like.
- Point out some right angles in the classroom. For those we can reach, how could we check?
- Which of these shapes has a right angle?

What to do for those children who achieve *above* expectation

- Ask the children to use a set-square to draw right angles and to compare angles with a right angle. (Level 3)

What to do for those children who achieve *below* expectation

- When asking the children to recognise and use whole, half and quarter turns, both clockwise and anticlockwise, only give them one instruction at a time.
- When asking the children to identify right angles, use RCM 26: Shape cards instead of RCM 24: Shapes and solids.

Task 20

Measuring

Objective **NC AT 3** **NC Level 2**

- Estimate, compare and measure lengths, weights and capacities, choosing and using standard units (m, cm, kg, litre) and suitable measuring instruments

Resources

Length:

- about 6 objects measuring between 10 cm and 2 m
- 30 cm ruler
- metre stick marked in centimetres
- tape measure

Weight:

- about 6 objects weighing between 1 kg and 5 kg
- balance
- 1 kg, 2 kg and 5 kg weights
- scales

Capacity:

- about 6 empty containers of different shapes and sizes between 250 ml and 5 litres
- bucket of water
- 1 litre measuring jug
- funnel

- selection of uniform non-standard measures: counters, cubes, match sticks, marbles, etc. (for children achieving *below* expectation)

Task

NOTE: This task is best undertaken with only two or three children.

- Place the ruler, metre stick, tape measure, balance, weights, scales, bucket of water, measuring jug and funnel on the table in front of the children.

Success criterion: *Estimate, compare and measure lengths, choosing and using standard units (m, cm) and suitable measuring instruments*

- Show the children the objects measuring between 10 cm and 2 m.
- Ask individual children questions similar to the following:
  - **Which of these objects is the longest / shortest?**
  - **Which objects are longer / shorter than the ...?**
  - **Which objects are longer / shorter than 1 metre?**
  - **Tell me about the length of these 2 / 3 objects.**
  - **Compare the length of these 2 / 3 objects.**
  - **Tell me something that is longer / shorter than...**

- Referring to specific objects, ask children questions similar to the following:
  - **How long do you think the … is?**
  - **How could you find out how long it is?**
  - **What could you use to measure it?**
  - **What units of measure would be suitable / would not be suitable for measuring it?**
- Ask individual children to find the length of specific objects. Say: **Louise, measure the length of the …**
- Continue until each child has sufficiently demonstrated their ability to estimate, compare and measure lengths, choosing and using standard units (m, cm) and suitable measuring instruments.
- Place the length objects to one side.

Success criterion: *Estimate, compare and measure weights, choosing and using standard units (kg) and suitable measuring instruments*

- Show the children the objects weighing between 1 kg and 5 kg.
- Ask individual children questions similar to the following:
  - **Which of these objects is the heaviest / lightest?**
  - **Which objects are heavier / lighter than the …?**
  - **Which objects weigh more than / less than 1 kilogram?**
  - **Tell me about the weight of these 2 / 3 objects.**
  - **Compare the weight of these 2 / 3 objects.**
  - **Can you tell me something that is heavier / lighter than …?**
- Referring to specific objects, ask children questions similar to the following:
  - **How heavy do you think the … is?**
  - **How could you find out how heavy it is?**
  - **What could you use to measure it?**
  - **What units of measure would be suitable / would not be suitable for weighing it?**
- Ask individual children to find the weight of specific objects. Say: **Louise, can you weigh the … for me?**
- Continue until each child has sufficiently demonstrated their ability to estimate, compare and measure weights, choosing and using standard units (kg) and suitable measuring instruments.
- Place the weight objects to one side.

Success criterion: *Estimate, compare and measure capacities, choosing and using standard units (litre) and suitable measuring instruments*

- Show the children the empty containers of different shapes and sizes between 250 ml and 5 litres.
- Ask individual children questions similar to the following:
  - **Which of these containers can hold the most / least?**
  - **Which containers can hold more than / less than this container?**
  - **Which containers can hold more than / less than 1 litre?**
  - **Tell me about how much these 2 / 3 containers can hold.**
  - **Compare how much these 2 / 3 containers can hold.**
  - **Can you tell me something that can hold more than / less than this container?**
- Referring to specific containers, ask children questions similar to the following:
  - **How much do you think this container can hold?**
  - **How could you find out how much it can hold?**
  - **What could you use to find out?**
  - **What units of measure would be suitable / would not be suitable for finding out how much it can hold?**

- Ask individual children to find the capacity of specific containers. Say: **Jasmine, can you find out for me how much this container can hold?**
- Continue until each child has sufficiently demonstrated their ability to estimate, compare and measure capacities, choosing and using standard units (litres) and suitable measuring instruments.

AfL

- Show me something that you think is just shorter / longer than a metre. How could you check whether you are right?
- Should we measure the … in centimetres or metres? Why? Would it be better to measure with a tape measure or a ruler?
- What can you see that you think is just shorter / longer than a metre?
- Point out something that you think is about two metres high / tall / long.
- When you use a balance, how could you find out if something is heavier than a kilogram? What would you need to do?
- Tell me an object in the classroom that you think is heavier / lighter than a kilogram. How could you check if it is?
- Which containers do you think will hold just a little more than a litre?
- Do you think the bucket holds 5 litres of water? How can we find out?
- Finish these sentences: 'I can measure the length of the classroom in …' 'I can measure the capacity of this bucket in …'
- Think of something that would be better measured in metres rather than centimetres. Explain why.

What to do for those children who achieve *above* expectation

- Encourage the children to be more accurate in estimating, comparing and measuring different lengths, weights and capacities. They should use a greater range of standard units, including kilometres, metres and centimetres; kilograms and grams; and litres and millilitres. They should also be more accurate when suggesting suitable units and equipment for such measurements.

What to do for those children who achieve *below* expectation

- When estimating, comparing and measuring lengths, weights and capacities, allow children to use direct comparisons and uniform non-standard measures.

Task 21

Measuring

Objective **NC AT 3** **NC Level 2**

- Read the numbered divisions on a scale, and interpret the divisions between them, e.g. on a scale from 0 to 25 with intervals of 1 shown but only the divisions 0, 5, 10, 15 and 20 numbered; use a ruler to draw and measure lines to the nearest centimetre

Resources

- several copies of RCM 28: Scales (enlarged to A3)
- sheet of paper showing several lines drawn to the nearest centimetre (per child), e.g.

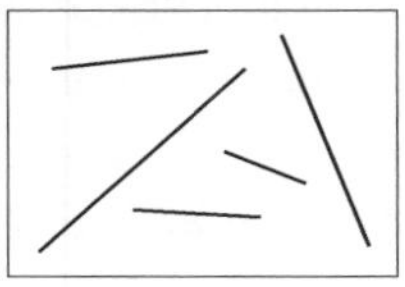

- ruler (per child)
- pencil and paper (per child)

Task

NOTES:

Prior to the task, mark different lengths, weights and capacities on the various scales on RCM 28, e.g.

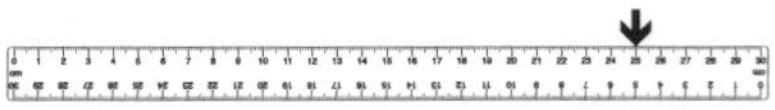

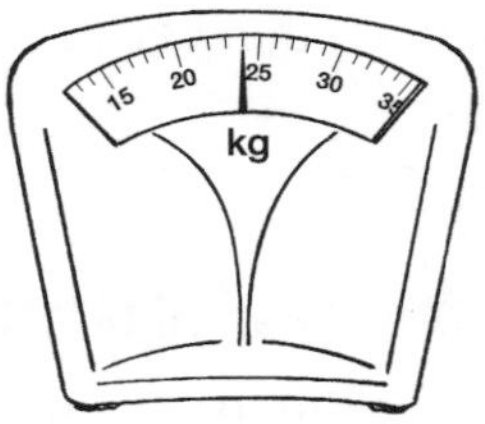

You may wish to mark up more than one copy of RCM 28 for this task. You will also need several blank copies of RCM 28.

Vary the markings on the scales according to the ability of the children undertaking in the task, i.e.

– mark a numbered division – easy
– mark an unnumbered division – moderate
– mark between two numbered or unnumbered divisions – difficult.

Success criterion: *Read the numbered divisions on a scale, and interpret the divisions between them – Length*

- Referring to the different rulers on RCM 28, ask questions that require the children to interpret the different readings. Point and say: **What length is the arrow pointing to on this ruler?**
- Mark a second point on the ruler and ask: **What is the distance between these two arrows?**
- Referring to a blank copy of RCM 28, ask individual children to record a length on one of the rulers. Point and say: **Yolanda, show me where 28 cm is on this ruler.**
- Continue until each child has sufficiently demonstrated their ability to read the numbered divisions on a scale, and interpret the divisions between them – length.

Success criterion: *Read the numbered divisions on a scale, and interpret the divisions between them – Weight*

- Referring to the different weighing scales on RCM 28, ask questions that require the children to interpret the different readings. Point and say: **What weight is the arrow pointing to?**
- Referring to a blank copy of RCM 28, ask individual children to record a weight on one of the scales. Point and say: **Allan, if something weighed 27 kilograms, what would this look like on the scales?**
- Continue until each child has sufficiently demonstrated their ability to read the numbered divisions on a scale, and interpret the divisions between them – weight.

Success criterion: *Read the numbered divisions on a scale, and interpret the divisions between them – Capacity*

- Referring to the different containers on RCM 28, ask questions that require the children to interpret the different readings. Point and say: **What is the water level in this container?**
- Referring to a blank copy of RCM 28, ask individual children to record a capacity on one of the containers. Point and say: **Aaron, if this container had 800 millilitres of liquid in it, what would this look like?**
- Continue until each child has sufficiently demonstrated their ability to read the numbered divisions on a scale, and interpret the divisions between them – capacity.

Success criterion: *Use a ruler to measure lines to the nearest centimetre*

- Provide each child with a sheet of paper showing several lines drawn to the nearest centimetre, a ruler and a pencil.
- Ask the children to use their ruler to measure the length of each of the lines of their sheet of paper and to write the length next to the line.
- Occasionally point and ask: **Yolanda, how long is this line? How much longer / shorter is this line than that line?**
- Repeat the above until each child has sufficiently demonstrated their ability to measure lines to the nearest centimetre.

Success criterion: *Use a ruler to draw lines to the nearest centimetre*

- Ensure that each child has a ruler, a pencil and a piece of paper.
- Ask individual children to draw lines of various lengths to the nearest centimetre, e.g. **Yolanda on your sheet of paper, I want you to draw a line 8 centimetres long. Aaron, I want you to draw a line 16 centimetres long.**
- Repeat the above until each child has sufficiently demonstrated their ability to draw lines to the nearest centimetre.

AfL

- What length is this ruler showing?
- What is the distance between these two points?
- What weight is this set of scales showing?
- Show me a weight of 150 g.
- How much liquid is in this container?
- Show me what this container would look like if it had 450 ml of liquid in it.

What to do for those children who achieve *above* expectation

- Vary the markings on the scales according to the ability of the children undertaking in the task, i.e. mark between the numbered or unnumbered divisions.
- Ask the children to draw lines to the nearest half centimetre or millimetre. (Level 3)

What to do for those children who achieve *below* expectation

- Vary the markings on the scales according to the ability of the children undertaking in the task, i.e. mark a numbered division.

Task 22

Measuring

Objective **NC AT 3** **NC Level 2**

- **Use units of time (seconds, minutes, hours, days) and know the relationships between them; read the time to the quarter hour; identify time intervals, including those that cross the hour**

Resources

- RCM 29: Time dominoes 1 (cut out)
- RCM 30: Time dominoes 2 (cut out)
- 2 demonstration analogue clocks
- 2 demonstration 12-hour digital clocks
- large sheet of paper and marker
- pencil and paper (per child)

Task

Success criterion: *Know units of time*

- Ask children questions similar to the following:
  - **How many minutes are there in an hour?**
  - **How many seconds in a minute?**
  - **How many days in March?**
  - **How many days are there in a year?**
  - **How many seasons are there?**

Success criterion: *Suggest suitable units of time*

- Ask children questions similar to the following:
  - **How long does it take you to walk to school / change for PE / eat your lunch...?**
  - **How long do you spend at school each day / asleep each night / doing maths each day...?**
  - **How old are you / your brothers and sisters / your parents / your grandparents...?**
  - **What unit of measurement would you use to measure the length of a television programme / the growing of a bean plant / the length of a sneeze / when you might become a grandparent...?**

Success criterion: *Know the relationship between units of time*

- Explain to the children that they are going to play a game of 'Time dominoes'.
- Spread the dominoes from RCMs 29 and 30 face down on the table.
- Share out all the dominoes amongst the children in the group, i.e.

Number of children in group	2	3	4	5	6
Number of dominoes per child	18	12	9	7	6

- If there is one domino left over, (i.e. group of 5 children) place the domino in the middle of the table and start with this. Otherwise, choose a child to start.

- Explain the rules of the game to the children.
  - The child who starts places a domino in the middle of the table.
  - The child to their left places a domino against the first, so that touching sides match:
  - Continue the play in rotation.
  - Unlike conventional dominoes where you can only join a domino at either end of the chain, allow the children to place a domino at any point along the chain as long as only one side of the domino is in contact with another domino and the values match (see right). (Note: there are four matching values for 2 days, 1 month, 2 months and 3 months, and six matching values for 2 weeks.)
  - If a child cannot go they miss a turn.
  - Continue the game until all the dominoes have been used.
  - As the children play the game, assess each child's ability to know the relationships between units of time.

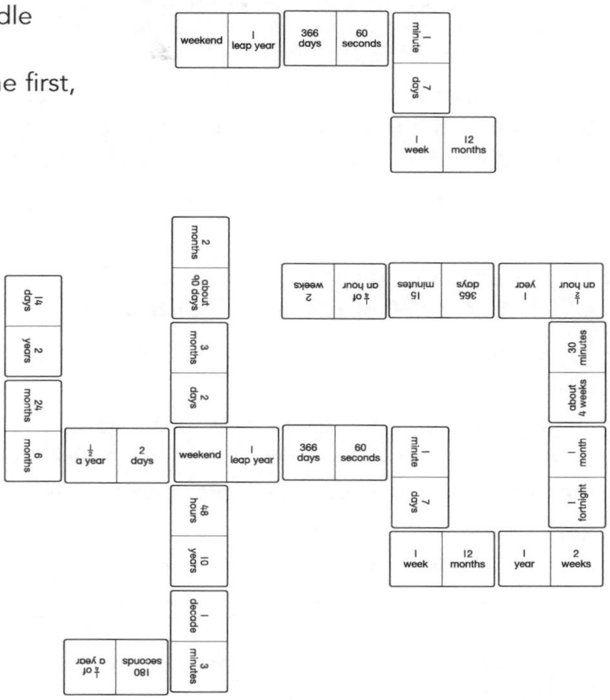

Success criterion: *Read the time to the quarter hour on an analogue clock*

- Show a time to the quarter hour on one of the analogue clocks, e.g. 7:45. Ask: **Harriet, what time does this clock show?** (7:45) **How else could you say this time?** (quarter to 8)
- Repeat several times until each child has sufficiently demonstrated their ability to read the time to the quarter hour on an analogue clock.

Success criterion: *Display the time to the quarter hour on an analogue clock*

- Give a child one of the analogue clocks and say: **Gita, show me quarter past 11 on this clock.**
- Repeat several times until each child has sufficiently demonstrated their ability to display the time to the quarter hour on an analogue clock.

Success criterion: *Read the time to the quarter hour on a 12-hour digital clock*

- On the large sheet of paper, write a time to the quarter hour using 12-hour digital clock notation, e.g. 12:15. Ask: **Gita, what time is this?** (12:15) **How else could you say this time?** (quarter past 12)
- Repeat several times until each child has sufficiently demonstrated their ability to read time to the quarter hour on a 12-hour digital clock.

Success criterion: *Display the time to the quarter hour on a 12-hour digital clock*

- Provide each child with a pencil and a piece of paper.
- Show a time to the quarter hour on one of the analogue clocks, e.g. 2:15. Say: **Look at this clock. I want each of you to write down what time this would show on a digital clock.**
- Repeat several times until each child has sufficiently demonstrated their ability to display the time to the quarter hour on a 12-hour digital clock.

Success criterion: *Identify time intervals, using an analogue clock and a 12-hour digital clock, including those that cross the hour*

- Show a different time, to the quarter hour, on each of the analogue clocks, e.g. 9:15 and 11:45. Say: **Both these times are in the morning**. Referring to each clock in turn, ask: **What time does this clock show? And this clock? What is the difference in time between quarter past 9 and quarter to 12?**
- Repeat the above several times, asking the children to calculate time intervals using two analogue clocks.
- Then show a different time, to the nearest quarter hour, on each of the 12-hour digital clocks, e.g. 7:45 and 8:15. Ask: **What is the time difference between these two clocks?**
- Repeat the above several times, asking the children to calculate time intervals using two 12-hour digital clocks.
- Next, show one time, to the quarter hour, on one of the analogue clocks and another quarter hour time on one of the 12-hour digital clocks, e.g. 10:15 and 12:00. Pointing to each clock in turn, ask: **How much time has passed from this time to this time?**
- Repeat the above several times, asking the children to identify time intervals using an analogue clock and a 12-hour digital clock, including those that cross the hour.

AfL

- Why did you place that domino there?
- Could you put the domino somewhere else?
- What time does this analogue clock read? Show me the same time using this digital clock.
- What time does this digital clock read? Show me the same time using this analogue clock.
- What is the difference in time between these two clocks?
- What time did the journey begin? How did you work it out?

What to do for those children who achieve *above* expectation

- Choose a domino and lay it out in front of a child. The child has to say as many things as they can think of that are related to that time, e.g.

7 days
5 working days plus 1 weekend

1 week	12 months

365 days
1 year
52 weeks

- Ask the children to read the time to the nearest five minutes and use a.m. and p.m. (Level 3)
- Ask questions that involve calculating time intervals using both analogue and digital clocks and that also cross the p.m. and a.m. boundary, e.g. 8:35 p.m. and 1:10 a.m.

What to do for those children who achieve *below* expectation

- Only use the 20 dominoes on RCM 29. You need to have groups of 2, 4 or 5 children for this task. Share out all the dominoes amongst the children in the group, i.e.

Number of children in group	2	4	5
Number of dominoes per child	10	5	4

  – Play the game as above. (Note: in this set of dominoes there are four matching values for each of 2 weeks and 1 month.)
- Ask the children to read time to the nearest hour and half hour.
- Ask questions that involve calculating time intervals using either analogue or digital clocks and that do not cross the hour, e.g. 11:15 and 11:45.

Task 23

Handling data
Using and applying mathematics

Objectives **NC AT 2 & 1** **NC Level 2**

- Answer a question by collecting and recording data in lists and tables; represent the data as block graphs or pictograms to show results; use ICT to organise and present data
- Follow a line of enquiry; answer questions by choosing and using suitable equipment and selecting, organising and presenting information in lists, tables and simple diagrams

Resources

- RCM 31: Collecting, recording and presenting data (per child)
- squared paper (per child)
- pencil (per child)
- ruler (per child)
- ICT data handling package – optional (per child or group)

Task

- Prior to the task, write a question in the box at the top of RCM 31. Choose a topic to investigate that is relevant to your particular circumstances and of interest to the children, e.g. What is Year 2's favourite colour? How old are children in Year 2? Most of the children in Year 2 have their birthday in summer. True or false? If you roll two dice and add the two numbers together, what is the most common total? What time do most children in Year 2 wake up in the morning? In which country were most Year 2 parents born? Alternatively, you may wish the children to suggest their own line of enquiry.

NOTE: You may want the children to work in pairs for this task.

Success criteria: *Identify what data to collect*
Collect data
Record data in lists and tables
Present data in block graphs or pictograms
Analyse and interpret data

- Provide each child with a copy of RCM 31, some squared paper, a pencil and a ruler.
- Briefly discuss the question with the children.
- Discuss with the children the first four questions on the RCM:
  - What information do you need to collect?
  - How are you going to collect the information?
  - How are you going to record the information?
  - How are you going to present the information? Why?
- Ask the children to collect, record and present the data.
- When the children have done this, they write about what they found out and what things they would do differently and the same if they were asked to answer the questions again.

AfL

- What information will you need to collect to answer these questions? How will you collect it? How will you display your data?
- What does this graph tell you? What makes the information easy or difficult to interpret?
- Make up three questions that can be answered using the data in this table/graph/chart.
- What further information could you collect to answer the question more fully?
- Which tables/charts/graphs are easy/difficult to interpret information from? Why?
- How did you work out that answer? Which information in the table/chart/graph did you use?

What to do for those children who achieve *above* expectation

- Ask the children to present their data using an ICT data handling package.

What to do for those children who achieve *below* expectation

- Ask the children to work in pairs.

Task 24

Handling data

Objective **NC AT 2** **NC Level 2**

- **Use lists, tables and diagrams to sort objects; explain choices using appropriate language, including 'not'**

Resources

- different classroom resources ideal for sorting, e.g. playing cards, 1–100 number cards, counters, interlocking cubes, beads, set of polygons, set of 3-D solids, paint brushes, coloured pencils, crayons, Compare Bears
- pencil and paper (per child)

Task

NOTES: This task is best undertaken with only two or three children. Decide whether it is appropriate to ask the children to begin sorting objects using one criterion (easy) or more than one criterion (difficult).

Success criteria: *Use lists, tables and diagrams to sort objects according to given criteria*
Explain choices using appropriate language, including 'not'

- Referring to the different classroom resources ideal for sorting, ask the children to sort the resources according to a given criterion, e.g.

Resource:	playing cards	1–100 number cards	counters / interlocking cubes / beads	set of polygons / set of 3-D solids
One criterion (easy):	– red cards / black cards – picture cards / not picture cards – numbers less than 5 / numbers 5 or more	– odd numbers / even numbers – numbers 1–50 / numbers 51–100 – one-digit numbers / two-digit numbers	– red / not red – cubes / spheres	– 2-D shapes / 3-D solids – green shapes / yellow shapes – circles / not circles – shapes with 4 sides or fewer / shapes with more than 4 sides
More than one criterion (difficult):	– sort according to the different suits – sort according to the card number	– multiples of 2 / multiples of 5 / not multiples of 2 or 5 – numbers 1–33 / numbers 34–66 / numbers 67–99	– counters / cubes / beads – blue / red / not blue or red – sort according to colour – cubes / spheres / not cubes or spheres	– blue shapes / red shapes / not blue or red shapes – shapes with fewer than 4 sides / shapes with 4 sides / shapes with more than 4 sides

- Once the children have sorted the resource, ask them to make a list or table. Say: **Look at the cards you have just sorted.** Ask questions such as: **How could you display the objects you have just sorted in a list? ...table? Which would be better: a list or a table? Why? How many columns are you going to need in your table? What headings are you going to have?**
- Continue until each child has sufficiently demonstrated their ability to use lists, tables and diagrams to sort objects according to given criteria, explaining choices using appropriate language, including 'not'.

Success criterion: *Use lists, tables and diagrams to sort objects according to their own criterion*

- After the children have sorted several different resources in different ways according to a given criterion (involving both one criterion and more than one criterion), ask the children to sort the resources according to their own criterion. Ask: **How else might you sort these Compare Bears? ...playing cards? ...number cards?**
- Once the children have sorted the resources, ask: **How have you sorted these objects?**
- Then, once again, ask the children to make a list or table of the sorted objects.
- Continue until each child has sufficiently demonstrated their ability to use lists, tables and diagrams to sort objects according to their own criterion, explaining choices using appropriate language, including 'not'.

AfL

- Where would you put these objects on this diagram? Why?
- One of the numbers in this diagram is wrong. Which number is it? Where does it belong?
- Explain to me how you have sorted these objects.
- How did you sort the objects / numbers?
- Why have you placed this object in this set?
- These objects have been sorted into two sets. How do you think they have been sorted?
- Why have you put this object in this part of the diagram?
- What else could be placed here?
- Why is this object not in the other set?

What to do for those children who achieve *above* expectation

- Ask the children to use Venn diagrams or Carroll diagrams to sort data and objects using more than one criterion. (Level 3)

What to do for those children who achieve *below* expectation

- Ask the children to sort objects according to a given criterion.

Self assessment Unit A1

Name ______________________ Date ____________

- I can explain to others how I solved a problem ☺ 😐 ☹
- I can read and write numbers to 100 ☺ 😐 ☹
- I know which numbers are odd and which are even ☺ 😐 ☹
- I can continue a number pattern ☺ 😐 ☹
- I can count objects by putting them into groups ☺ 😐 ☹
- I know what each digit in a number less than 100 stands for ☺ 😐 ☹
- I can write numbers in order and position them on a number line ☺ 😐 ☹
- I can use the greater than (>) and less than (<) symbols to show that one number is larger or smaller than another ☺ 😐 ☹
- I can estimate a group of objects ☺ 😐 ☹
- I can add and subtract some numbers in my head ☺ 😐 ☹
- I know that addition and subtraction 'undo' each other ☺ 😐 ☹
- I can write three other related number sentences for $6 + 3 = 9$ ☺ 😐 ☹

Self assessment Unit B1

Name ______________________ Date ____________

- I can continue a number pattern ☺ 😐 ☹
- I can explain to others how I solve a problem ☺ 😐 ☹
- I can solve problems involving money ☺ 😐 ☹
- I can recall addition and subtraction number facts for each number up to 10 ☺ 😐 ☹
- I can count in steps of 2, 5 or 10 ☺ 😐 ☹
- I can double and halve numbers ☺ 😐 ☹
- I know that if I double a number then halve the answer I get back to the number I started with ☺ 😐 ☹
- I can check the answer to an addition by doing a related subtraction ☺ 😐 ☹
- I can look at pictures of 2-D shapes and name them ☺ 😐 ☹
- I can sort a set of 3-D solids ☺ 😐 ☹
- I know the order of the days of the week ☺ 😐 ☹

Self assessment Unit C1

Name ______________________ Date ____________

- I can decide what information I need to answer a question ☺ 😐 ☹
- I can put information in lists or tables ☺ 😐 ☹
- I know how to collect information ☺ 😐 ☹
- I can use lists and tables to show what I found out ☺ 😐 ☹
- I can sort objects and talk about how I sorted them ☺ 😐 ☹
- I can read numbers on a scale ☺ 😐 ☹
- I can find out if something is longer or shorter than a metre ☺ 😐 ☹
- I can use a ruler to measure in centimetres ☺ 😐 ☹

Self assessment Unit D1

Name ______________________ Date ___________

- I can decide what calculation to do to solve a problem ☺ 😐 ☹
- I can add some one-digit and two-digit numbers in my head ☺ 😐 ☹
- I can subtract some one-digit and two-digit numbers in my head ☺ 😐 ☹
- I can read numbers on a scale ☺ 😐 ☹
- I can use a ruler to measure and draw lines to the nearest centimetre ☺ 😐 ☹
- I can estimate how long an activity might take, then check using a timer ☺ 😐 ☹
- I can tell the time when it is something o'clock or half past the hour ☺ 😐 ☹
- I can follow and give instructions to mark a position on a grid ☺ 😐 ☹

Self assessment Unit E1

Name ______________________ Date ____________

- I know what information I need to use to solve a problem and can describe what I did step by step ☺ 😐 ☹
- I can record a calculation in a number sentence and check if my answer makes sense ☺ 😐 ☹
- I can use a number line to do multiplication and division ☺ 😐 ☹
- I know how to write number sentences for multiplication and division as well as addition and subtraction ☺ 😐 ☹
- I can explain what my number sentence means ☺ 😐 ☹
- I know doubles of numbers up to 10 and I can use what I know to work out halves ☺ 😐 ☹
- I understand the connection between doubling and halving ☺ 😐 ☹
- I can recognise some of the 2, 5 and 10 times tables and can explain the patterns I see ☺ 😐 ☹
- I can use these patterns to see if other numbers belong to the sequence ☺ 😐 ☹
- I can use my knowledge of halving numbers to help me to work out half and a quarter of a set of objects or a shape ☺ 😐 ☹

Self assessment Unit A2

Name ______________________ Date ____________

- I can explain how I solved a problem and say why I did it that way ☺ 😐 ☹
- I can count on and back in steps of 2, 5 and 10 ☺ 😐 ☹
- I can read and write numbers up to 100 in figures and in words ☺ 😐 ☹
- I know which numbers are odd and which are even ☺ 😐 ☹
- I can explain what each digit in a two-digit number stands for ☺ 😐 ☹
- I can partition numbers in different ways ☺ 😐 ☹
- I can add and subtract some numbers in my head ☺ 😐 ☹
- I can add and subtract bigger numbers using practical equipment or by writing notes to help me ☺ 😐 ☹
- I know how to write number sentences using the +, − and = symbols ☺ 😐 ☹
- I can explain what different number sentences mean ☺ 😐 ☹

Self assessment Unit B2

Name ______________________ Date ____________

I can decide which calculations to do to solve a problem	☺	😐	☹
I can recall addition and subtraction number facts for each number up to 10	☺	😐	☹
I know which pairs of numbers make 20	☺	😐	☹
I know which pairs of multiples of 10 make 100	☺	😐	☹
I know some of the number facts in the 2, 5 and 10 times tables	☺	😐	☹
I know that multiples of 2 are even numbers	☺	😐	☹
I can describe and continue number patterns	☺	😐	☹
I can read and write two-digit and three-digit numbers in words and figures	☺	😐	☹
I can draw a line of symmetry on a shape or picture	☺	😐	☹
I can complete a symmetrical picture by drawing the 'other half'	☺	😐	☹
I can make a symmetrical pattern using coloured tiles	☺	😐	☹
I can name and describe common 2-D shapes	☺	😐	☹

Self assessment Unit C2

Name ______________________ Date ___________

- I can organise information and make lists and tables ☺ 😐 ☹

- I can sort objects and use diagrams to show how I sorted them ☺ 😐 ☹

- I can make block graphs to show information ☺ 😐 ☹

- I understand information that is presented in tables, sorting diagrams and block graphs ☺ 😐 ☹

- I can estimate whether an object is heavier or lighter than a half-kilogram by putting a half-kilogram in one hand and the object in the other ☺ 😐 ☹

- I can read numbers on a scale and work out the numbers between them ☺ 😐 ☹

Self assessment Unit D2

Name ______________________ Date ____________

- I can decide what calculation to do to solve a problem ☺ 😐 ☹
- I can solve problems involving pounds and pence ☺ 😐 ☹
- I can add more than two numbers in my head ☺ 😐 ☹
- I can add and subtract pairs of multiples of 10 ☺ 😐 ☹
- I can add and subtract larger numbers using practical equipment or notes to help me ☺ 😐 ☹
- I can read numbers on a scale and can work out the numbers between them ☺ 😐 ☹
- I can solve problems involving weight ☺ 😐 ☹
- I can tell the time when it is quarter past, half past or quarter to the hour ☺ 😐 ☹
- I know that a quarter past three is the same time as three fifteen ☺ 😐 ☹
- I can order events in time ☺ 😐 ☹
- I can turn on the spot through whole, half or quarter turns, either clockwise or anticlockwise ☺ 😐 ☹
- I can follow and give directions ☺ 😐 ☹

Self assessment Unit E2

Name ______________________ Date ____________

• I know what I need to do to help me solve a problem and then I can work out the answer	☺	😐	☹
• I can show how I solved a problem and explain steps in my working	☺	😐	☹
• I can use calculations to solve problems and I know which calculation to use	☺	😐	☹
• I can use sharing to work out divisions and can explain what I did	☺	😐	☹
• I know how to write number sentences for multiplication and for division	☺	😐	☹
• I can explain what different number sentences mean	☺	😐	☹
• I know some of my doubles up to 20	☺	😐	☹
• I know some of my times tables for 2, 5 and 10	☺	😐	☹
• I can use counting or other strategies for those times table facts I don't know	☺	😐	☹
• I can recognise some multiples of 2, 5 and 10	☺	😐	☹
• I can fold a piece of paper into halves or quarters	☺	😐	☹
• I can find a half or a quarter of a set of objects	☺	😐	☹

Self assessment Unit A3

Name ______________________ Date ____________

- I can show and explain clearly how I solved a problem ☺ 😐 ☹
- I can explain the pattern for a sequence of numbers and work out the next few numbers in the list ☺ 😐 ☹
- I can write numbers in order and position them on a number line ☺ 😐 ☹
- I can round a two-digit number to the nearest 10 ☺ 😐 ☹
- I can use the greater than (>) and less than (<) symbols to show that one number is larger or smaller than another ☺ 😐 ☹
- I can say roughly how many there are in a group of objects ☺ 😐 ☹
- I can add more than two numbers ☺ 😐 ☹
- I can add and subtract two-digit numbers using practical equipment or written notes to help me ☺ 😐 ☹
- I know when it is easier to use addition to work out a subtraction ☺ 😐 ☹
- I can write two addition number sentences and two subtraction number sentences using the numbers 2, 3 and 5 ☺ 😐 ☹

Self assessment Unit B3

Name ______________________ Date ____________

- I can describe and continue the pattern for a set of numbers or shapes ☺ 😐 ☹
- I can decide which calculations are needed to solve a two-step word problem ☺ 😐 ☹
- I know which pairs of numbers make 20 ☺ 😐 ☹
- I know all the pairs of multiples of 10 that make 100 ☺ 😐 ☹
- I know the doubles of all the numbers up to 20 and the matching halves ☺ 😐 ☹
- I know my 2, 5 and 10 times tables and can work out the division facts that go with them ☺ 😐 ☹
- I can tell if a number is a multiple of 2, 5 or 10 ☺ 😐 ☹
- I can check answers to calculations involving doubling by halving the answer ☺ 😐 ☹
- I can name and describe familiar 2-D shapes ☺ 😐 ☹
- I can name and describe familiar 3-D solids ☺ 😐 ☹
- I can solve problems involving time and capacity ☺ 😐 ☹

Self assessment Unit C3

Name ______________________________ Date ______________

- I can test out an idea by collecting and organising information ☺ 😐 ☹
- I can sort objects and use diagrams to show how I sorted them ☺ 😐 ☹
- I can draw block graphs to show information ☺ 😐 ☹
- I can draw pictograms to show information ☺ 😐 ☹
- I understand information that is presented in tables, sorting diagrams, pictograms and block graphs ☺ 😐 ☹
- I can compare two or more containers to see which holds more ☺ 😐 ☹
- I can use a measuring jug to measure a litre of water and to find out how much water other containers hold ☺ 😐 ☹
- I can read scales marked in 1s, 2s, 5s and 10s ☺ 😐 ☹

Self assessment Unit D3

Name ____________________________ Date ______________

• I can decide which calculations are needed to solve a two-step word problem	☺	😐	☹
• I can add and subtract two-digit numbers using practical equipment or written notes to help me	☺	😐	☹
• I can recognise patterns in different number sentences	☺	😐	☹
• I can solve problems involving time	☺	😐	☹
• I know that a quarter turn makes a right angle	☺	😐	☹
• I can point out right angles in the classroom	☺	😐	☹
• I can use quarter and half turns to make patterns	☺	😐	☹
• I know that a litre is 1000 millilitres	☺	😐	☹
• I can read scales marked in 100 ml measures	☺	😐	☹

Self assessment Unit E3

Name ______________________ Date ____________

- When I have worked out the answer to a problem, I can look again at the problem and then check that the answer makes sense ☺ 😐 ☹
- I can explain how I worked out the answer to a problem and can show the working I did ☺ 😐 ☹
- I can use arrays to help me work out multiplication ☺ 😐 ☹
- I can do multiplication and division in different ways and show how I do them ☺ 😐 ☹
- I can work out the missing numbers in number sentences ☺ 😐 ☹
- When I think I have the answer, I can put it in the number sentence and check whether it is correct ☺ 😐 ☹
- I can double all numbers up to 20 and can find matching halves ☺ 😐 ☹
- I know my 2, 5 and 10 times tables ☺ 😐 ☹
- I can work out divisions that go with the tables ☺ 😐 ☹
- I understand the idea of remainders when dividing ☺ 😐 ☹
- I can find three quarters of a set of objects or of a shape ☺ 😐 ☹

Test 1

Papers A and B answers

Paper A

1. 55
2. 20
3. rectangle, square
4. 12
5. 

6. 16
7.

Round to 10	Round to 20	Round to 30
6	23 17	32 28

8. 23 − 8 = 15 (or 23 = 8 + 15)
9. 35p
10. 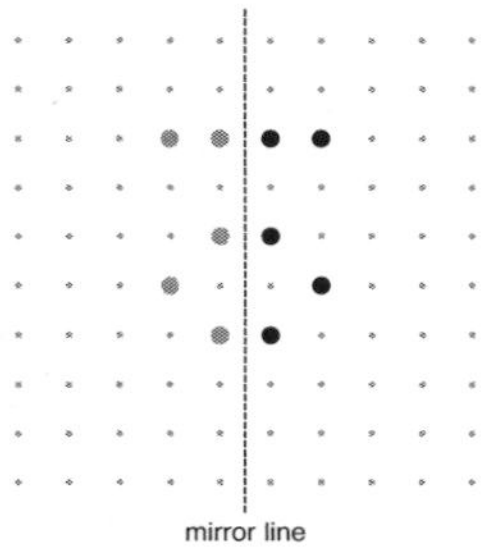

11. 18, 45, 50, 61, 73, 82
12. 20
13. August
14. 65
15. 2 cm
16. 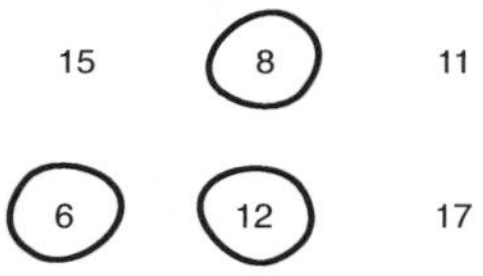

17. 6
18. 10:15
19. 0, 7
20. 3
21. 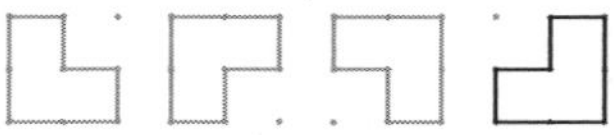
22. 5 + 3 = 8, 3 + 5 = 8, 8 − 5 = 3, 8 − 3 = 5
23. 29, 26
24. 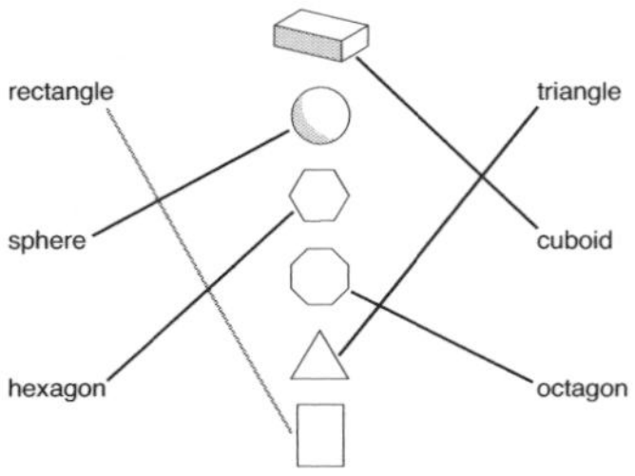

25. *Check that the child has shaded 4 more squares.*
26. 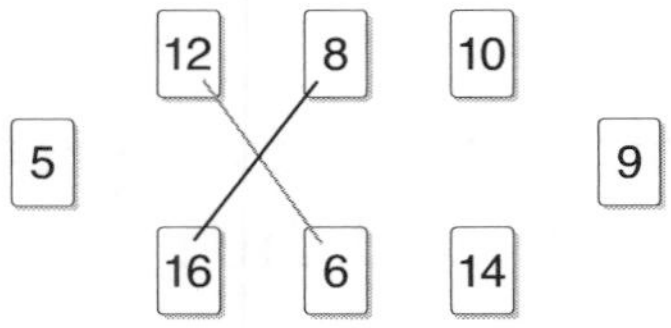

27. *Check that the length of the line is 11 cm.*
28. 4 → 8
    3 → 6
    1 → 2
    8 → 16
29. 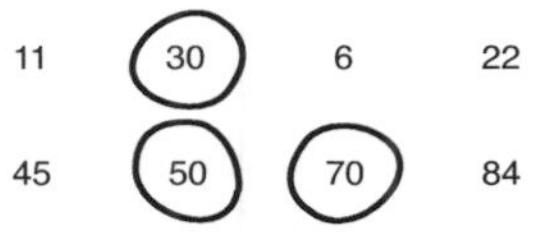

30. 

Paper B

1. 74
2. 48 years old
3. 2
4. 340
5. 2 m 80 cm
6. 30
7. *Accept any of the following numbers:* 154, 194, 514, 594, 914 or 954
8. 20p and £1.25
9. 18
10. 9
11. 75
12. 382
13. 

14. 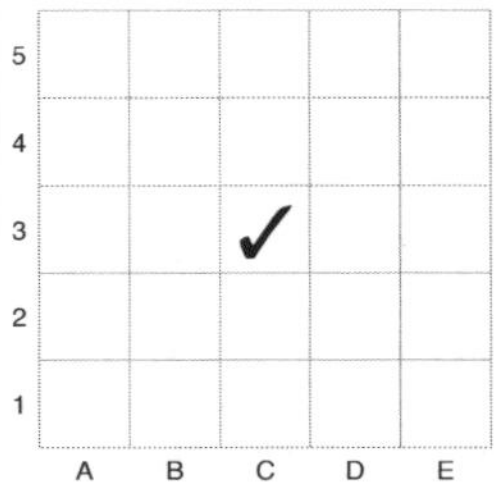

15. 26 cm
16. 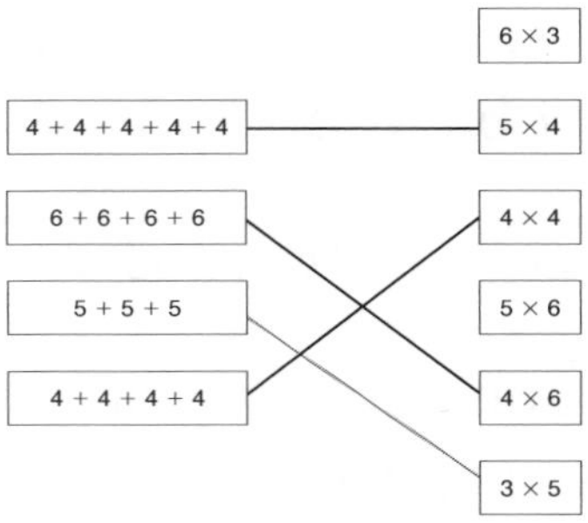

17. 1006
18. £3.65, £3.75, **£3.85**, £3.95, **£4.05**, £4.15
19. 107
20. 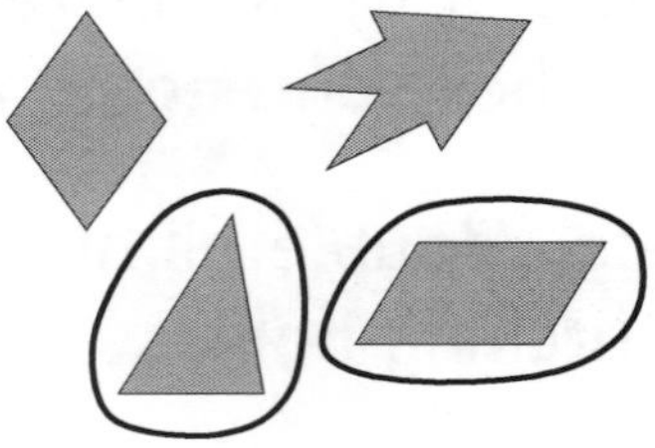
21. 17, **56**, 121, **206**, 236, **370**, 426
22. *Accept any two numbers in the boxes that when added to 270 equal 430, e.g.* 270 + **50** + **110** = 430
23. *Check child's method, e.g.*
    10 × 5 = 50
    8 × 5 = 40
    50 + 40 = 90
24. *Check child's shape has 8 sides.*
25. 25
26.

	is taller than 120 cm	is not taller than 120 cm
has blue eyes		**John**
does not have blue eyes	**Lucy**	

27. *Check child has shaded half the shape, i.e. 4 triangles.*
28. *Check that the child has drawn a line 13 cm long.*
29. 219
30. 12

Test 1A

Mental mathematics test questions

Say: **Listen carefully as I read aloud some questions for you to answer. I will read each question twice. The first question is a practice question which we will all do together. The boxes are for you to write your answers in. The letters below each box show you which box to use for each question.**

Question	Say
Practice	**Listen to this sequence: 10, 20, 30,** [one clap] **50, 60** **Write the missing number in box a.**
1	**Listen to this sequence: 15, 25, 35, 45,** [one clap] **65** **Write the missing number in box b.**
2	**What is the sum of 13 and 7?** **Write your answer in box c.**
3	**Look at the names of the shapes in box d.** **They say: triangle, rectangle, hexagon, pentagon, square.** **Circle the names of the shapes which have four sides.**
4	**There are 4 oranges in each pack.** **Mrs Rogers buys 3 packs of oranges.** **How many oranges does she buy?** **Write your answer in box e.**
5	**Look at the watch in box f.** **Rav leaves home at eight-fifteen.** **It takes him half an hour to get to work.** **Draw the hands on the watch that show what time Rav gets to work.**

Year 2

Mathematics booklet

TEST 1 PAPER A	LEVEL 2

Name

Date

Class

You need
- **pencil**
- **ruler**
- **mirror**

PAGE	MARKS	PAGE	MARKS
2		10	
3		11	
4		12	
5		13	
6		14	
7		15	
8		16	
9		**TOTAL** (out of 30)	

Practice question

a

1

b

1

1 mark

2

c

2

1 mark

Total out of 2

3

triangle rectangle

hexagon pentagon

square

d

3
1 mark

4

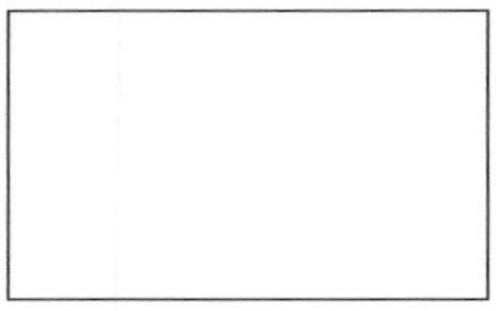

e

4
1 mark

5

f

5
1 mark

Practice question

Write the answer.

$5 + 4 = \square$

6

Write the answer.

$8 \times 2 = \square$

6
1 mark

7

Write each number in the correct box.
One has been done for you.

23 17 6 32 28

Round to 10	Round to 20	Round to 30
	23	

7
1 mark

8

Look at these signs.

$+ \quad - \quad \times \quad \div \quad =$

Write a sign in each box to make this correct.

23 ☐ 8 ☐ 15

8

1 mark

9

Mark and Elsa share this money equally.

How much do they each get? ☐ p

9

1 mark

10

Draw the reflection of this pattern in the mirror line.

You can use a mirror.

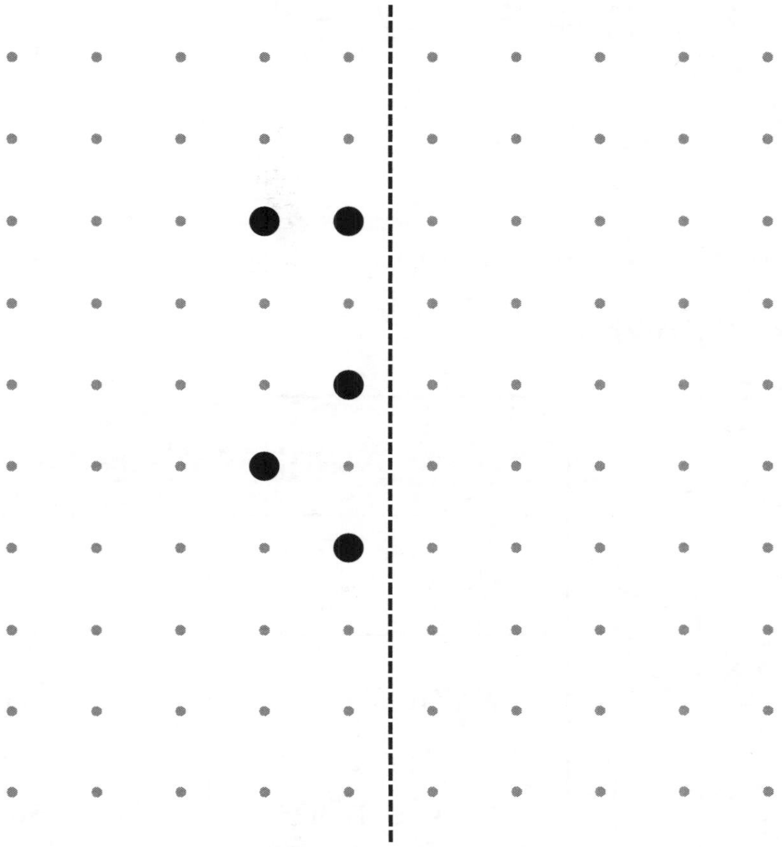

10
1 mark

11

Write these numbers in order, smallest to largest.

45 73 61 18 50 82

smallest largest

11
1 mark

12

Write the total.

$$4 + 3 + 6 + 7 = \square$$

12
1 mark

13

A shop sells bikes.

Number of bikes	
July	42
August	58
September	34
October	48
November	56
December	52

In which month did the shop sell the **most** bikes?

13
1 mark

14

Mark drew a number line to help him find the answer to 45 + 23.

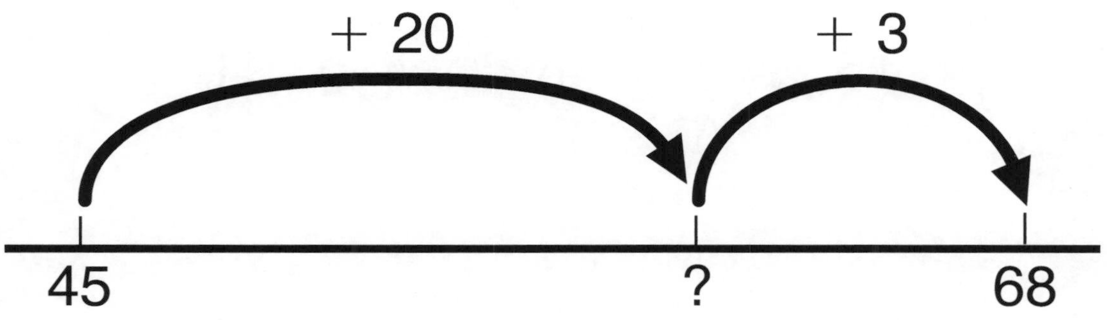

What number is missing?

14

1 mark

15

Measure these lines.

How much shorter is line A than line B?

15

1 mark

16

Circle the numbers that are **not** odd.

15 8 11

6 12 17

16
1 mark

17

Some children share 18 strawberries.

Each child gets 3 strawberries.

How many children are there?

17
1 mark

18

A bus left at this time to go to the museum.

It arrived 1 hour and 15 minutes later.

Circle the time it got to the museum.

18
1 mark

19

Look at this number line.

Write the two missing numbers in the boxes.

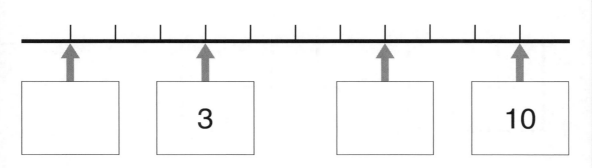

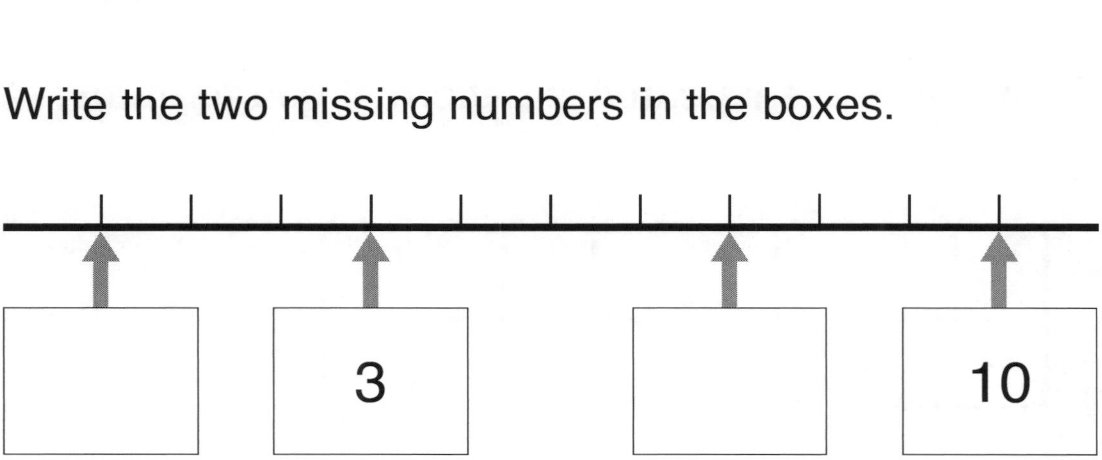

19
1 mark

20

Some children drew a pictogram.

Ice cream we like best

 stands for 1 child

chocolate	
strawberry	
toffee	
vanilla	

How many **more** children chose chocolate than vanilla?

20

1 mark

21

Draw the next shape in this pattern of quarter turns.

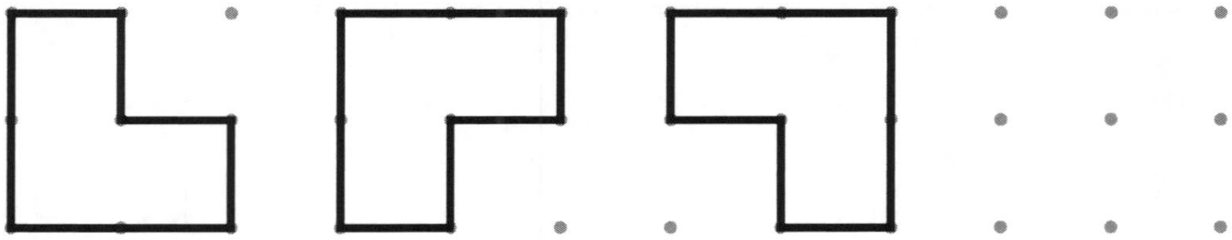

21

1 mark

22

Use these digit cards to write two addition number sentences and two subtraction number sentences.

8	5	3

☐ + ☐ = ☐

☐ + ☐ = ☐

☐ − ☐ = ☐

☐ − ☐ = ☐

22
1 mark

23

Write the missing numbers in the sequence.

38	35	32			23

23
1 mark

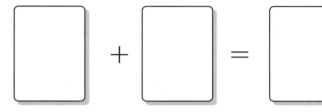

24

Draw lines to match the shape to its label.

One has been done for you.

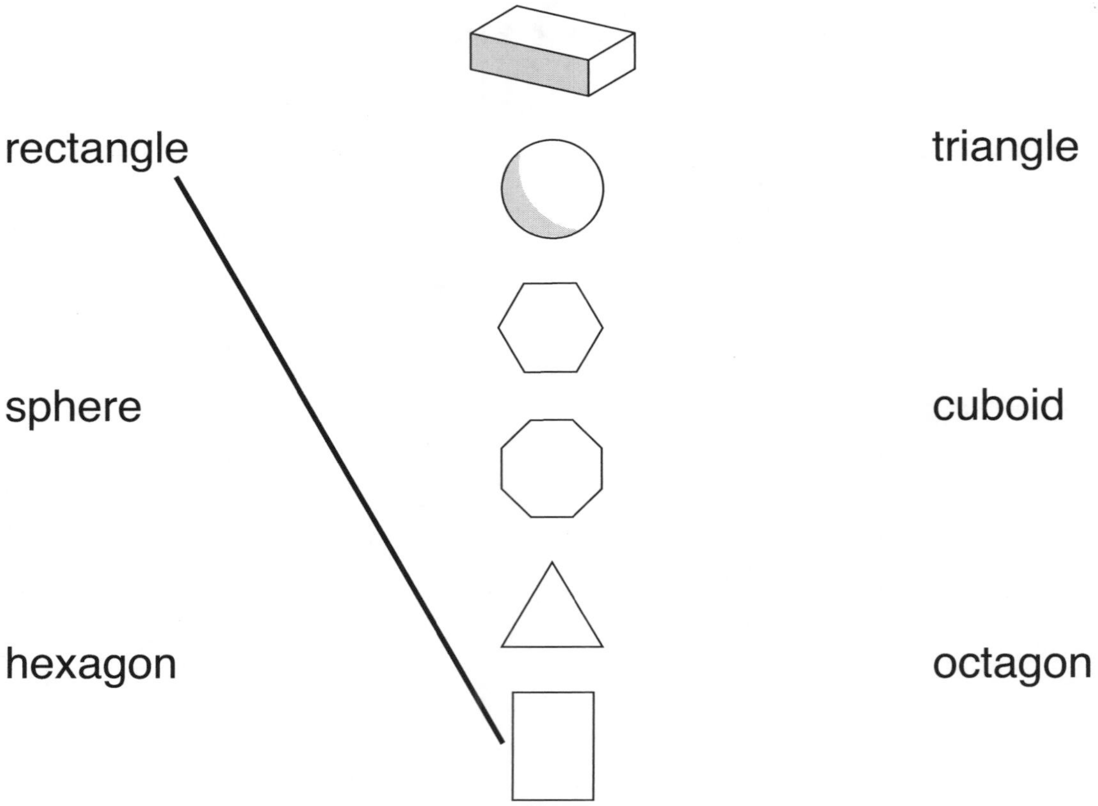

24

1 mark

25

Shade more squares so that exactly half of the shape is shaded.

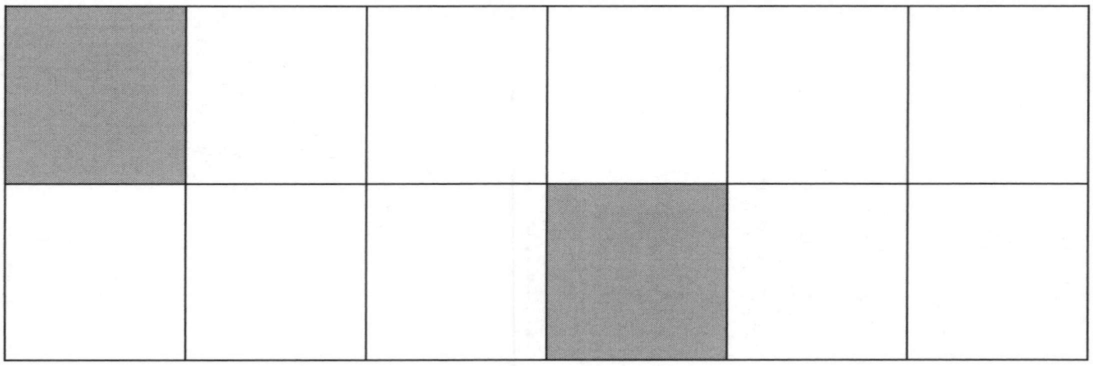

25

1 mark

26

The two numbers joined together have a difference of 6.

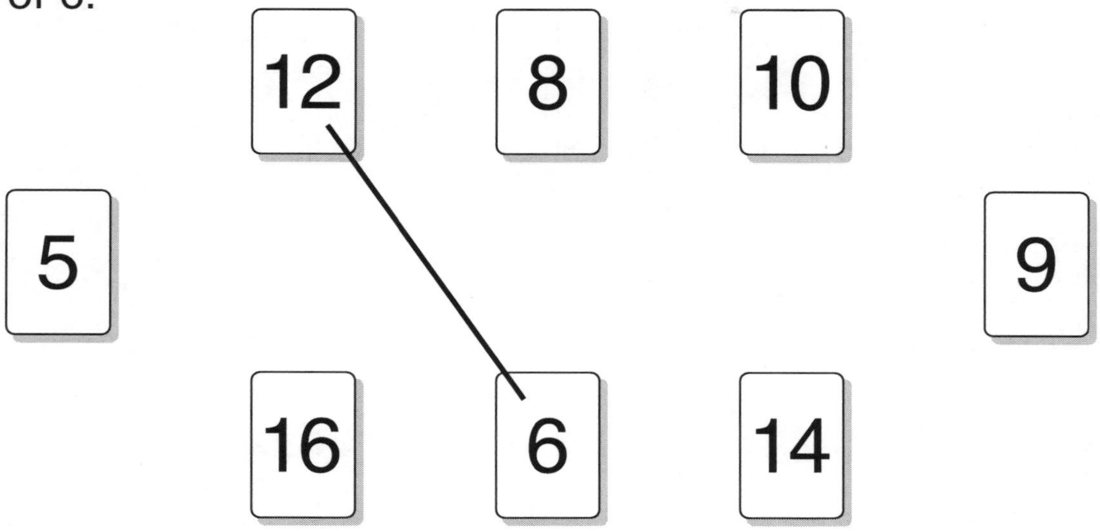

Join together two other numbers that have a difference of 8.

26

1 mark

27

Draw a line 11 centimetres long.

Use a ruler.

27

1 mark

28

Double each of these numbers.

4 ➔ [ ]

3 ➔ [ ]

1 ➔ [ ]

8 ➔ [ ]

28
1 mark

29

Some of these numbers can be divided exactly by 10.

11	30	6	22
45	50	70	84

Circle all the numbers which can be divided exactly by 10.

29
1 mark

30

The tally chart shows the number of different types of vehicles that passed a school one afternoon.

Vehicles

Vehicle	Tally	Total
bike	~~IIII~~ ~~IIII~~ II	12
bus	IIII	4
car	~~IIII~~ ~~IIII~~ ~~IIII~~ ~~IIII~~ III	23
truck	~~IIII~~ I	6
van		17

The tally for the vans is empty.

Complete the tally for the vans.

30

1 mark

End of test

Test 1B

Mental mathematics test questions

Say: **Listen carefully as I read aloud some questions for you to answer. I will read each question twice. The first question is a practice question which we will all do together. The boxes are for you to write your answers in. The letters below each box show you which box to use for each question.**

Question	Say
Practice	**Add 24 and 20.** **Write your answer in box a.**
1	**What is 100 subtract 26?** **Write your answer in box b.**
2	**Thomas is 23 years old.** **His father is 25 years older.** **How old is Thomas's father?** **Write your answer in box c.**
3	**Look at the shaded shape.** **How many right angles does it have?** **Write your answer in box d.**
4	**Look at box e.** **Write the number three hundred and forty in the box.**
5	**Look at the picture of the sunflower.** **Write the height of the sunflower in box f.**

Year 2

Mathematics booklet

TEST 1 PAPER B | LEVEL 3

Name

Date

Class

You need
- pencil
- ruler
- mirror

PAGE	MARKS	PAGE	MARKS
2		10	
3		11	
4		12	
5		13	
6		14	
7		15	
8		16	
9		**TOTAL** (out of 30)	

Practice question

a

1

b

1

1 mark

2

years old

c

2

1 mark

3

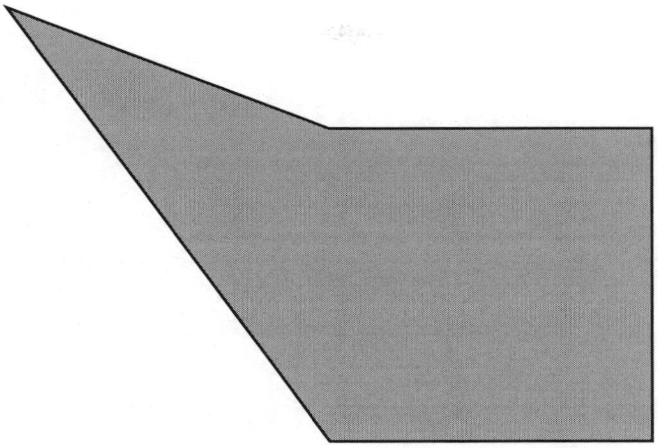

d

3

1 mark

4

e

4

1 mark

5

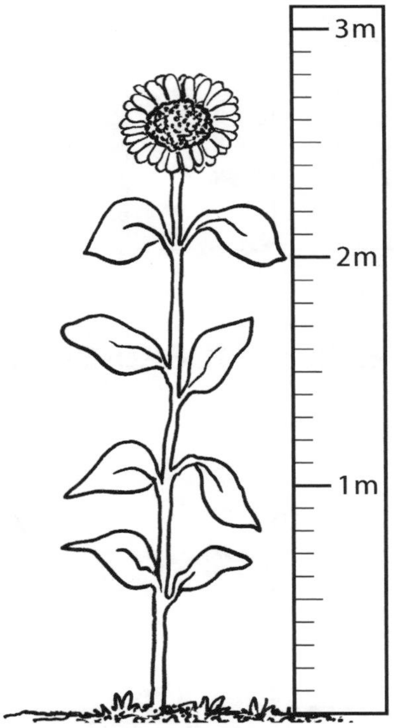

f

5

1 mark

Practice question

Write the answer.

$$5 + 4 = \square$$

6 Write the number in the box to make this correct.

$$80 - 30 = 20 + \square$$

6
1 mark

7

1	4	5	9

Choose three of these number cards to make an even number that is greater than 100.

7
1 mark

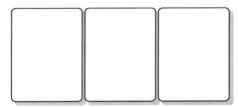

8

Here is a set of stamps.

Lucy posts a letter. It costs £1.45

She uses two of these stamps.

Which two stamps does she use?

[ ] and [ ]

8

1 mark

9

Write the number which is half of 36.

[ ]

9

1 mark

10

There are 45 children.

They get into teams of 5.

How many teams are there altogether?

10

1 mark

11

Estimate the number marked by the arrow.

Write the number in the empty box.

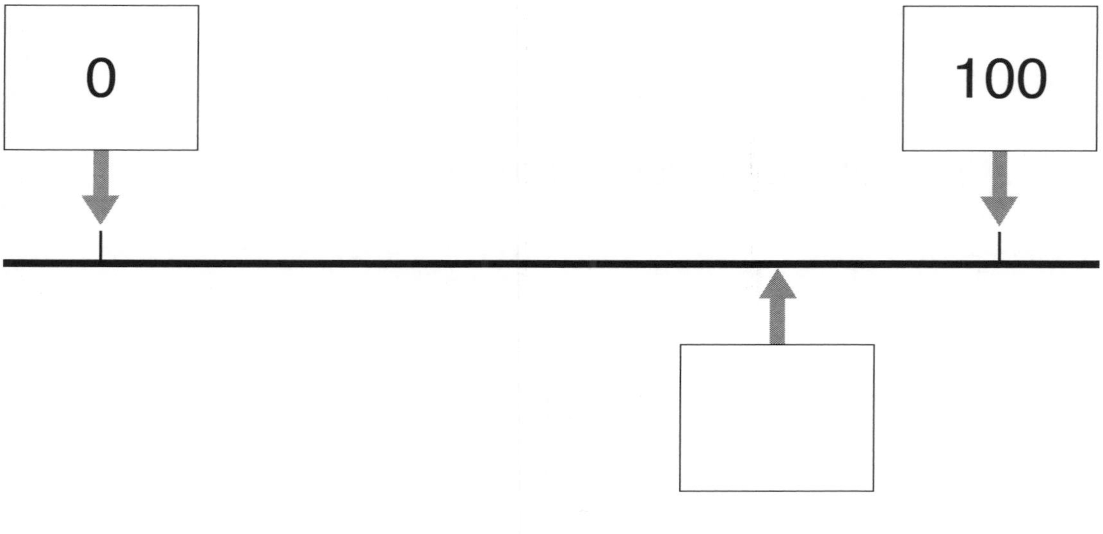

11

1 mark

12 Write the total.

$300 + 80 + 2 = \square$

12
1 mark

13 Two clocks show the same time.

Circle them.

13
1 mark

14

Tick (✓) the square that is exactly halfway between squares A1 and E5.

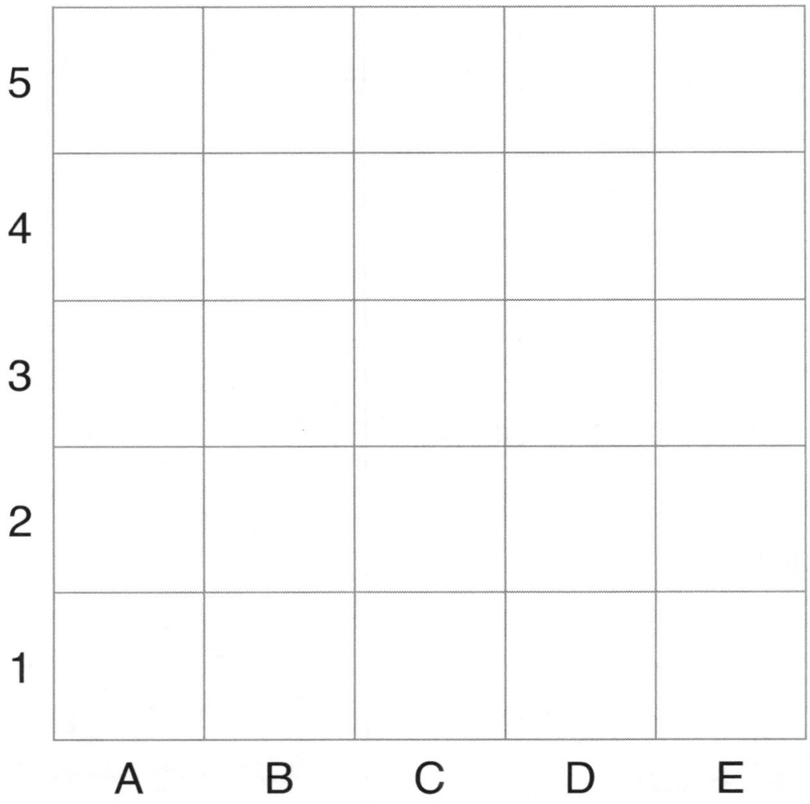

14
1 mark

15

Lucy is 134 cm tall.

John is 108 cm tall.

How much taller is Lucy than John?

15
1 mark

16

Match each addition to a multiplication.

One is done for you.

	6×3
$4 + 4 + 4 + 4 + 4$	5×4
$6 + 6 + 6 + 6$	4×4
$5 + 5 + 5$	5×6
$4 + 4 + 4 + 4$	4×6
	3×5

16

1 mark

17

Write the missing number.

996 —**10** more→ ☐

17

1 mark

18

Write the missing amounts in this sequence.

The same amount is added each time.

£3.65 £3.75 [ ] £3.95 [ ] £4.15

18

1 mark

19

A school held a concert.

283 people attended the concert.
92 were women.
84 were men.
The rest were children.

How many children attended the concert?

Show how you work it out.

[ ]

19

1 mark

20

Two of these shapes have no lines of symmetry.

Circle them.

You may use a mirror.

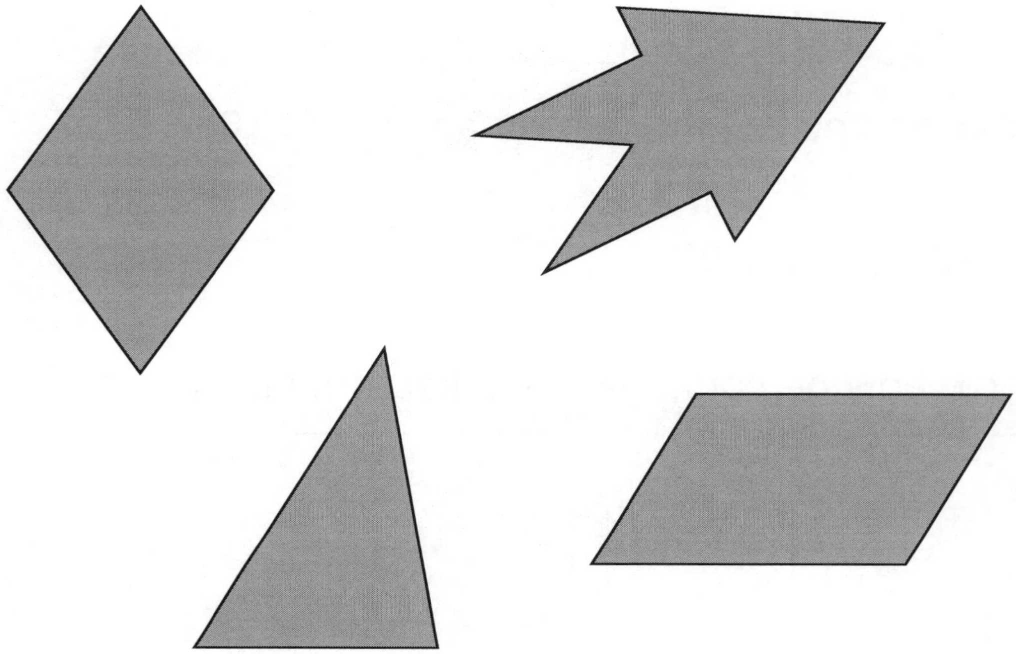

20

1 mark

21

Write these numbers on the cards.

All the numbers must be in order.

370 56 206

17		121		236		426

21

1 mark

22 Write a number in each box to make this correct.

270 + [ ] + [ ] = 430

22
1 mark

23 John worked out the correct answer to **18 × 5**

His answer was **90**

Show how he could have worked out his answer.

[ ] 90

23
1 mark

24

Use the dots to draw a different octagon.

You may use a ruler.

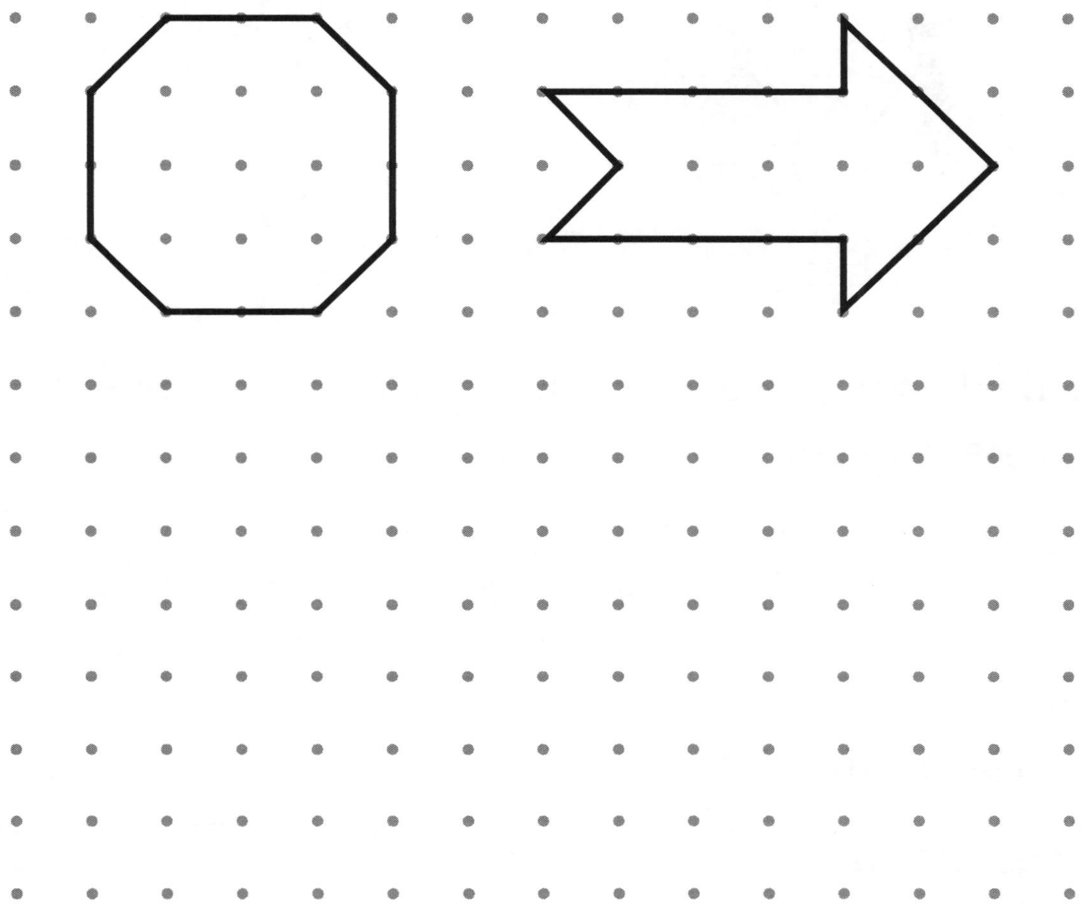

24

1 mark

25

Write the answer.

$73 - 48 = \square$

25

1 mark

26

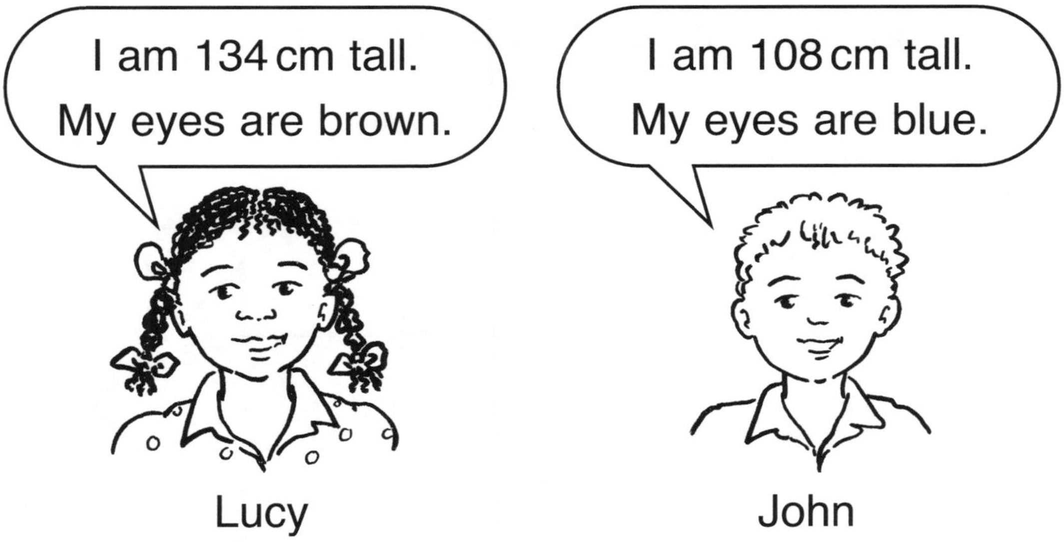
I am 134 cm tall.
My eyes are brown.
I am 108 cm tall.
My eyes are blue.
Lucy
John

Write Lucy's and John's names in the correct boxes on the diagram.

	is taller than 120 cm	is not taller than 120 cm
has blue eyes		
does not have blue eyes		

26

1 mark

27

Colour half of this shape.

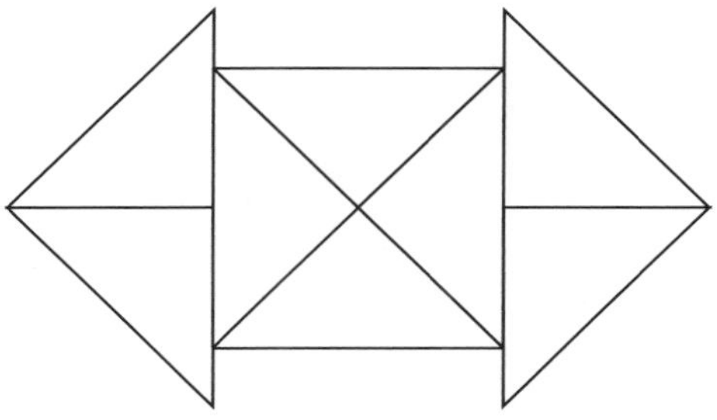

27

1 mark

Total out of 2

28 Draw one line which is twice as long as this line.

Use a ruler.

28
1 mark

29 Write the total.

$$147 + 72 = \square$$

29
1 mark

30

The tally chart shows the number of people that eat in a restaurant one week.

Number of customers

Day	**Tally**
Monday	C L O S E D
Tuesday	~~IIII~~ IIII
Wednesday	~~IIII~~ ~~IIII~~ ~~IIII~~ III
Thursday	~~IIII~~ ~~IIII~~ ~~IIII~~ ~~IIII~~ II
Friday	~~IIII~~ ~~IIII~~ ~~IIII~~ ~~IIII~~ ~~IIII~~ I
Saturday	~~IIII~~ ~~IIII~~ ~~IIII~~ ~~IIII~~ ~~IIII~~ III
Sunday	~~IIII~~ ~~IIII~~ IIII

How many more people eat at the restaurant on Friday than on Sunday?

30

1 mark

End of test

Test 2

Papers A and B answers

Paper A

1. 15
2. 20 years old
3. 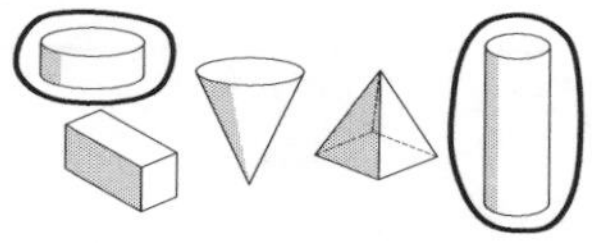
4. 26, 48, 71
5. 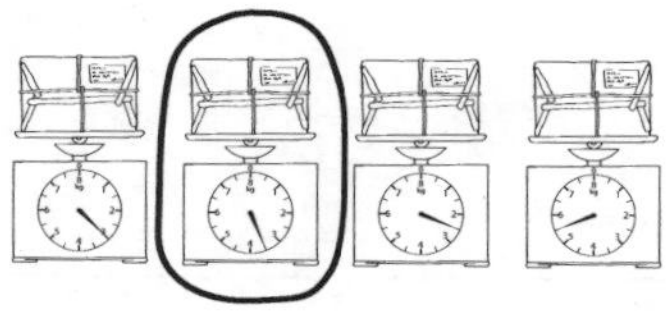
6. 25
7. >
8. 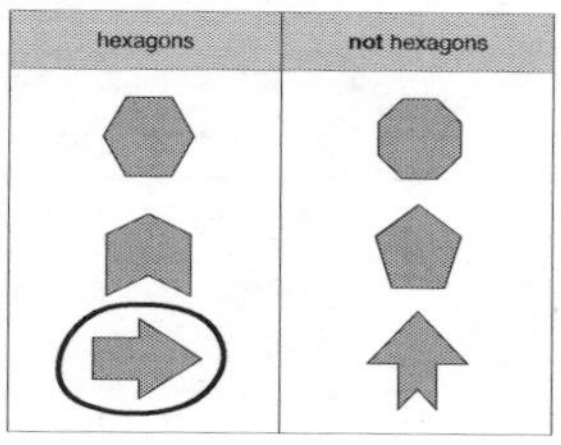

9. 63 64 57 68 (58)
10. 57
11. 21, 18, **15**, 12, 9, 6, **3**, 0
12. 25
13. 110
14. 142, 214, 241, 412, 421
15. 83
16. 
17. 6, 5
18. 14
19. 14
20. 8p
21. 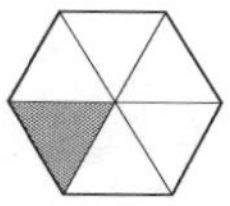
22. 5
23. 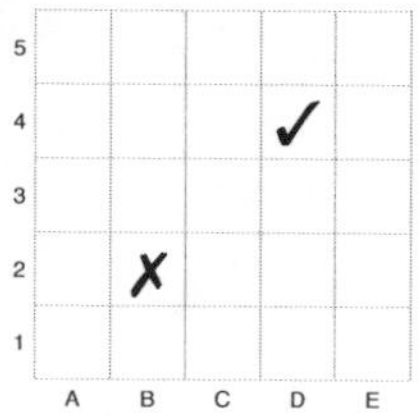

24. £2.87
25. 35
26. 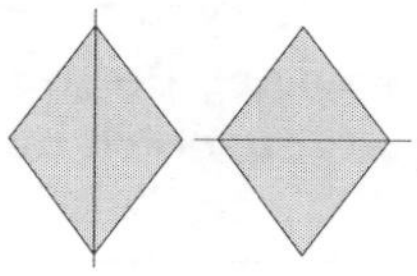
27. (35) (10) 56 18 (45) (30)
28. 7 + 5 + 6 = 18
29. 5 cm
30. 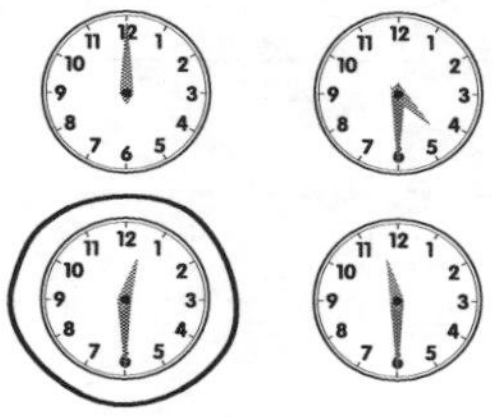

Paper B

1. 27
2. *Accept any two numbers with a product of 120, i.e.* 120 × 1, 60 × 2, 40 × 3, 30 × 4, 24 × 5, 20 × 6, 15 × 8, 12 × 10
3. 2
4. 81
5. $1\frac{1}{2}$ kg
6. 60, 20
7. 751
8. 35p
9. 92
10. 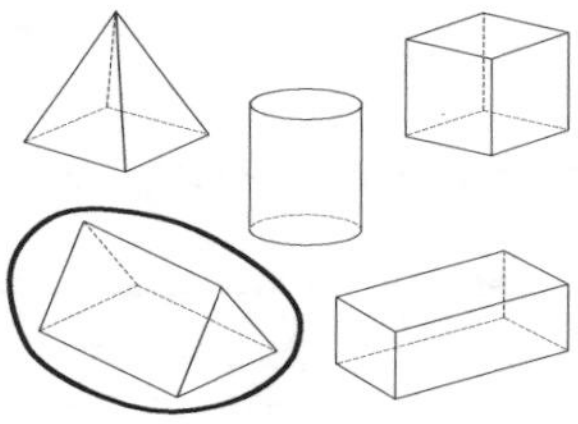
11. **14**, **17**, **20**, 23, 26, 29, **32**, **35**, **38**
12. 82
13. *Check numbers written in the Venn diagram.*
a = odd number > 100;
b = odd number < 100;
c = even number < 100
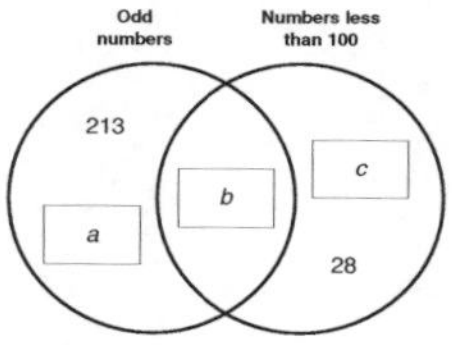

14. 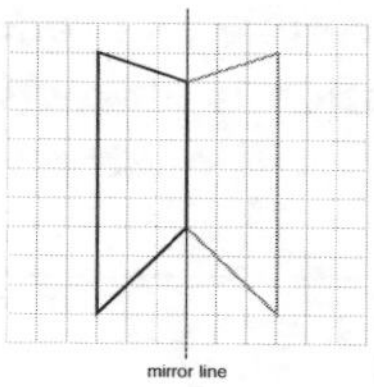

15. 225
16. 1000 g
17. 26
18. 160 g
19. 7
20. E5
21. *Check child's answer, e.g.* 6 → 12 → 6
22. 4:10
23. 39
24. *Accept either of the following:*
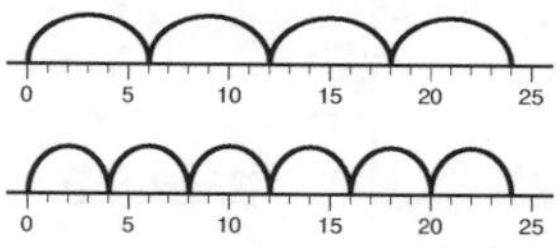

25. *Check that the child has shaded 4 more squares.*
26. 62, 126
27. 2 2 + 4 5 = 6 7
28. *Accept any two numbers that total a multiple of 10, i.e.* 11 and 19, 12 and 18, 13 and 17, 14 and 16.
29. 354
30. *Accept either of the following:*
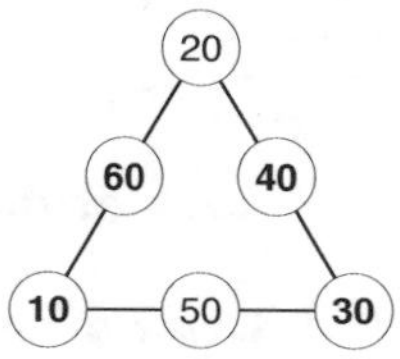

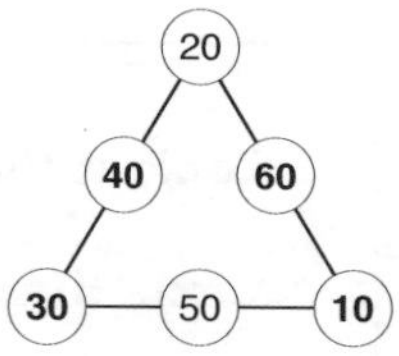

Test 2A

Mental mathematics test questions

Say: **Listen carefully as I read aloud some questions for you to answer. I will read each question twice. The first question is a practice question which we will all do together. The boxes are for you to write your answers in. The letters below each box show you which box to use for each question.**

Question	Say
Practice	**What number is 1 more than 8?** **Write your answer in box a.**
1	**Add these three numbers: 5, 5 and 5.** **Write your answer in box b.**
2	**Emma is 13 years old.** **Liam is 7 years older than Emma.** **How old is Liam?** **Write your answer in box c.**
3	**Look at the shapes.** **Circle each picture of a cylinder.**
4	**Look at the numbers.** **Draw a ring around these numbers: 26, 48 and 71.**
5	**Look at the four scales.** **Circle the scales which show $3\frac{1}{2}$ kilograms.**

Year 2

Mathematics booklet

TEST 2 PAPER A | **LEVEL 2**

Name

Date

Class

You need
- **pencil**
- **ruler**
- **mirror**

PAGE	MARKS	PAGE	MARKS
2		10	
3		11	
4		12	
5		13	
6		14	
7		15	
8		16	
9		**TOTAL** (out of 30)	

Practice question

a

1

b

1

1 mark

2

years old

c

2

1 mark

Total out of 2

3

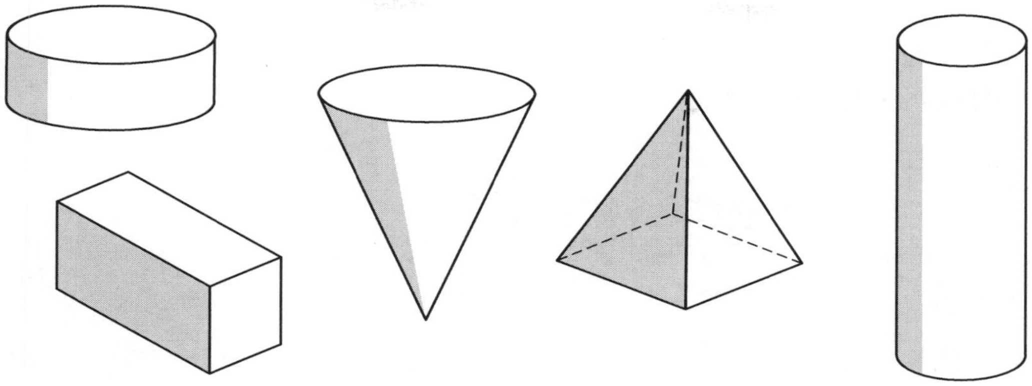

3

1 mark

4

58	26	66
21	48	17
57	62	71

4

1 mark

5

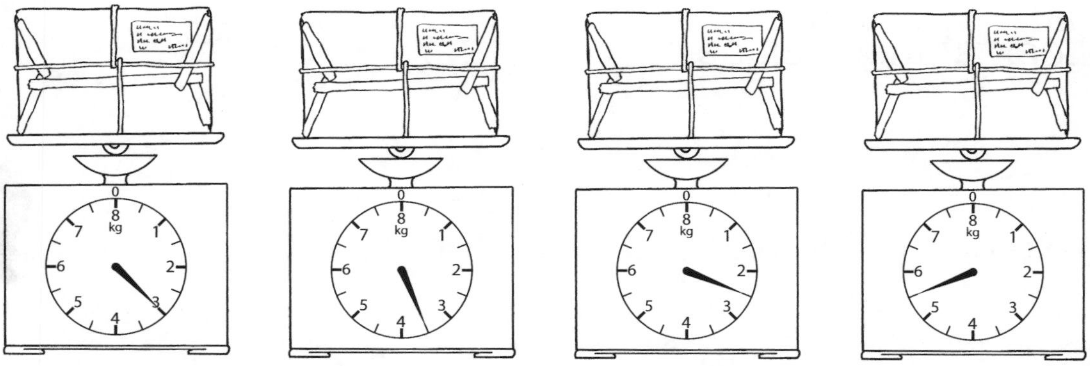

5

1 mark

Practice question

Write the answer.

$$4 \times 5 = \square$$

6 This number square is torn.

1	2	3	4	5
6	7	8	9	10
11	12	13	14	15
16	17	18	19	
21	22	23		

What was the largest number in the square?

6
1 mark

7 Look at these signs.

$<$ $>$ $=$

Use one of the signs to make this correct.

$$28 \quad \square \quad 17$$

7
1 mark

8

These shapes have all been sorted.

One shape is in the wrong place.

Circle it.

hexagons	**not** hexagons
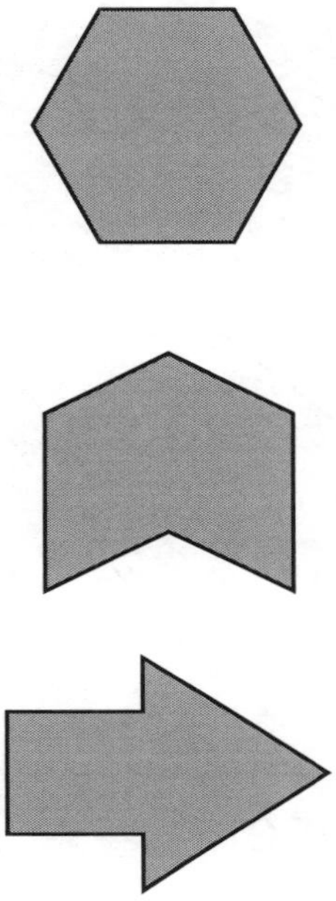	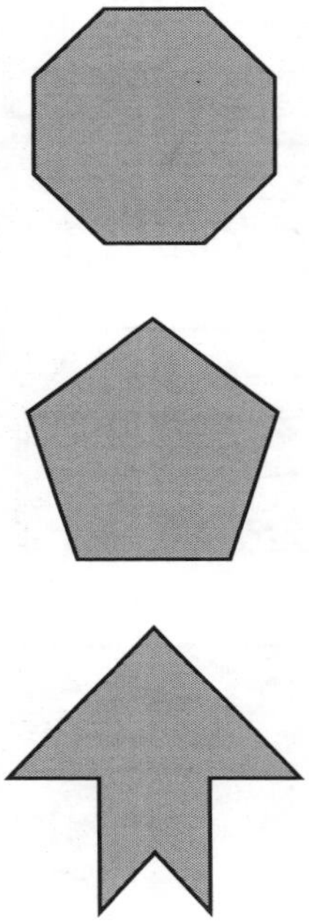

8

1 mark

9

Which number is nearest to 60?

Circle it.

63 64 57 68 58

9

1 mark

10

There are 10 pencils in each box and 7 more pencils.

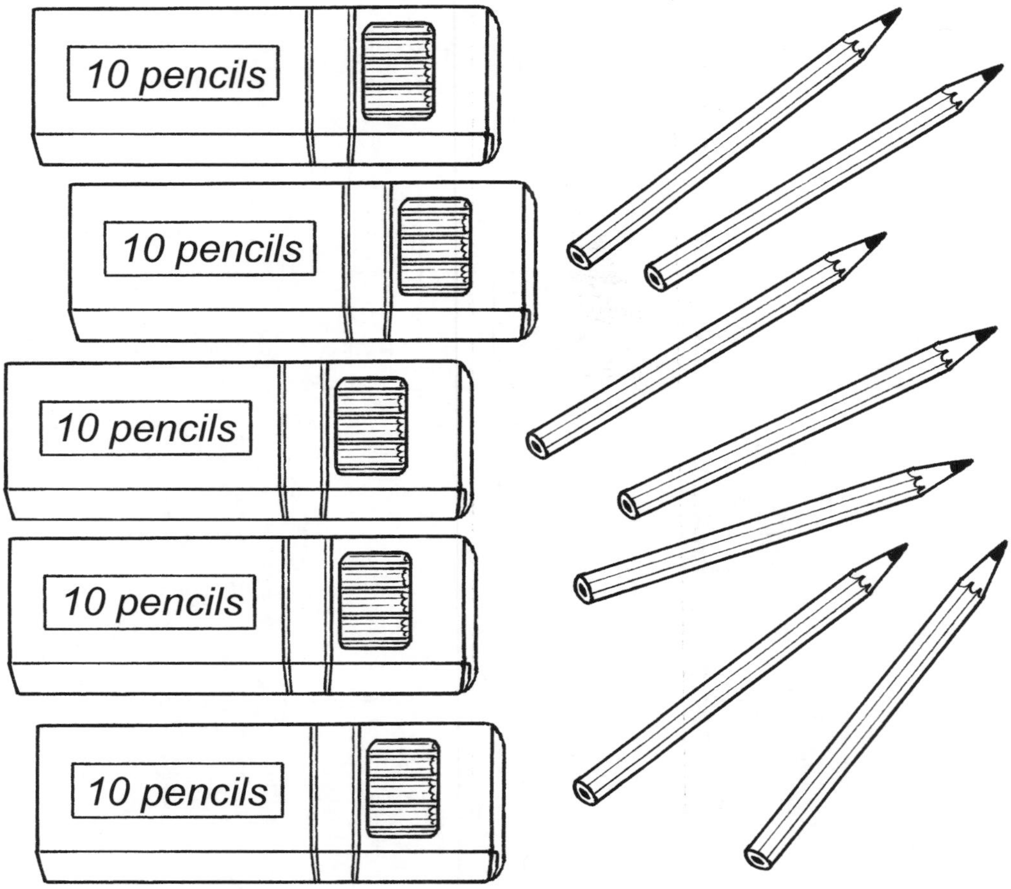

How many pencils are there altogether?

10

1 mark

11

Fill in the two missing numbers in this sequence.

21	18		12	9	6		0

11

1 mark

12

Year 2 drew a graph.

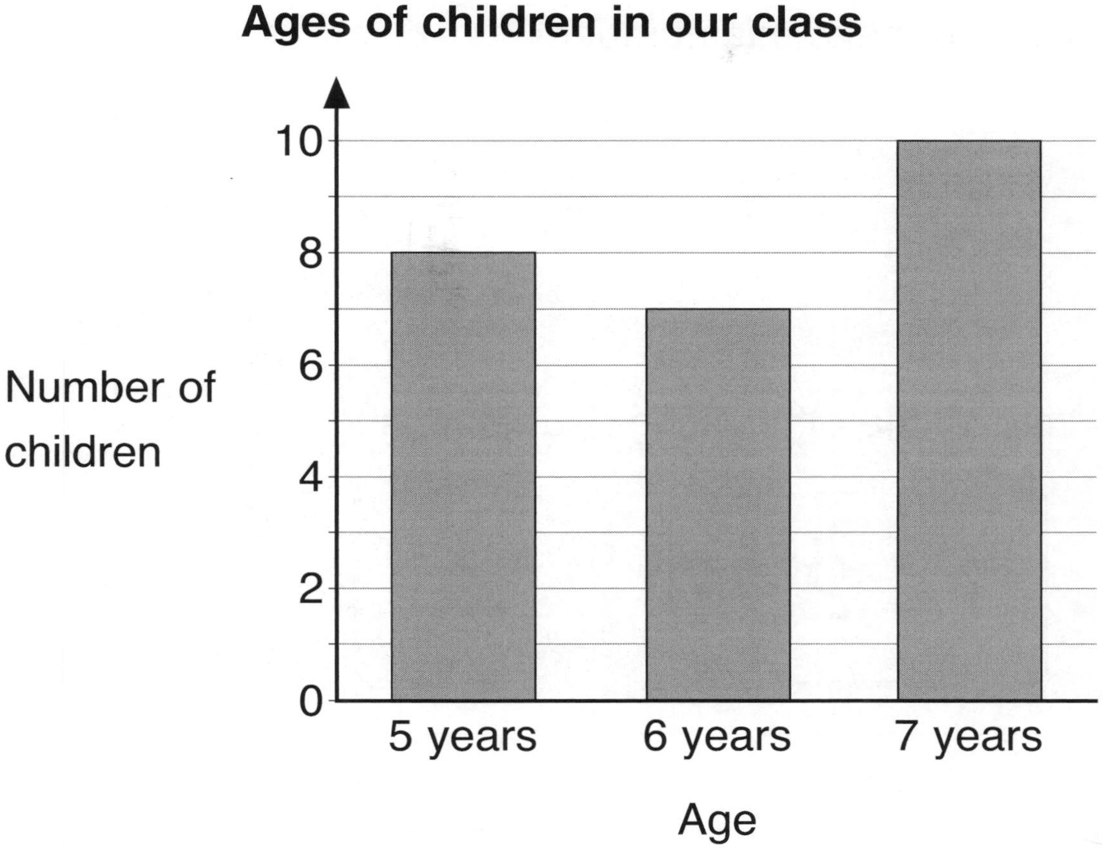

What is the total number of children in the class?

12
1 mark

13

Write the total.

$$20 + 50 + 30 + 10 = \square$$

13
1 mark

14

Lucy is making 3-digit numbers with these cards.

She can make this number.

1 2 4

Write all the other 3-digit numbers she can make.

1 2 4

14

1 mark

15

Write the total.

$26 + 57 =$

15

1 mark

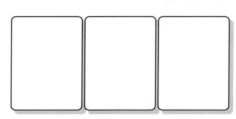

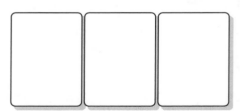

16

Draw an arrow on the dial to show the weight of sugar.

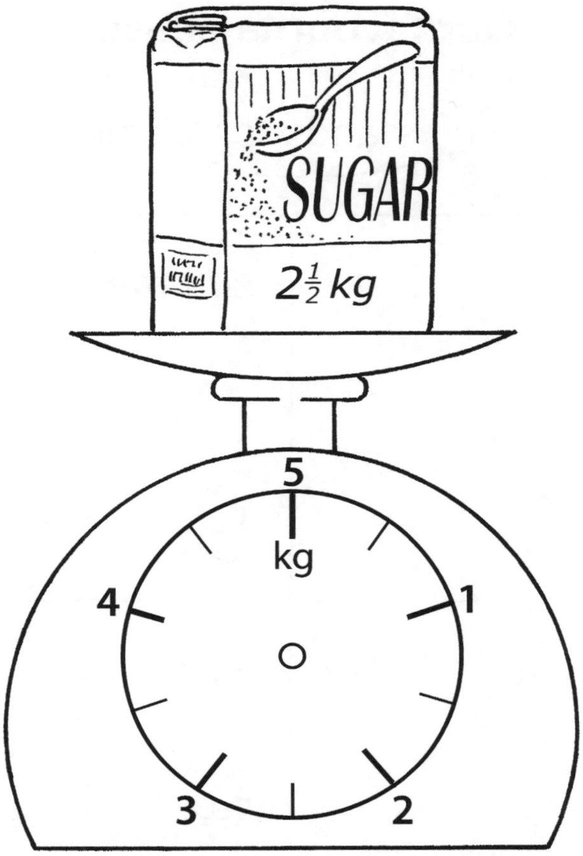

16

1 mark

17

Write numbers in the boxes to make these correct.

$9 - \square = 3$

$\square + 2 = 7$

17

1 mark

18

Some children drew a pictogram.

Our favourite season

 stands for 1 child

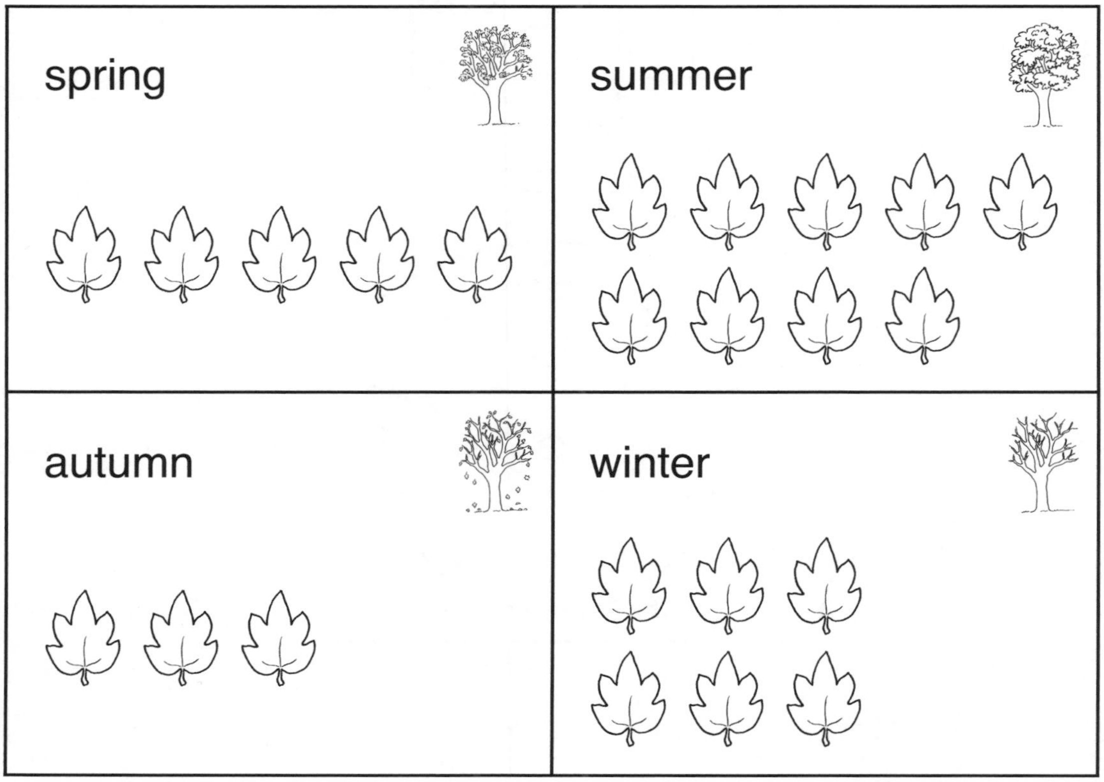

How many children prefer spring or summer?

18 1 mark

19

Write the answer.

$7 \times 2 = \square$

19 1 mark

20

Samir has 8p. He wants to buy an orange.

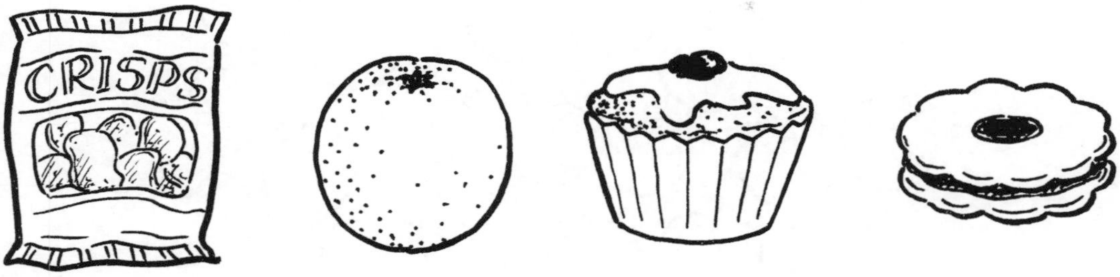

crisps	orange	cake	biscuit
24p	16p	15p	9p

How much more money does he need?

20

1 mark

21

Shade the correct triangle in the last hexagon.

starting shape	➔	after half a turn

21

1 mark

22

Jake eats one quarter of these apples.

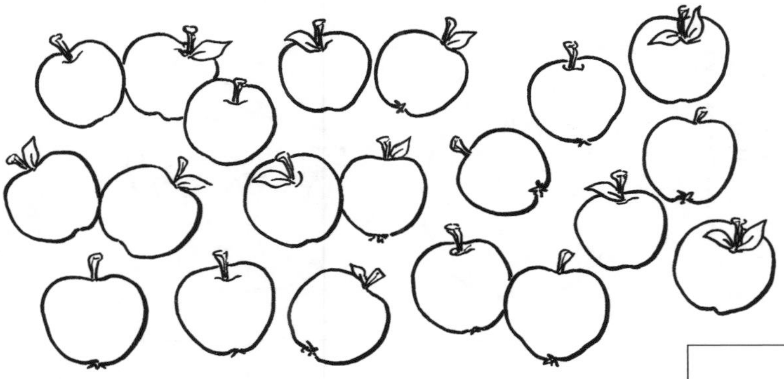

How many does he eat?

22

1 mark

23

The tick (✓) is in square D4.

Draw a cross (✗) in square B2.

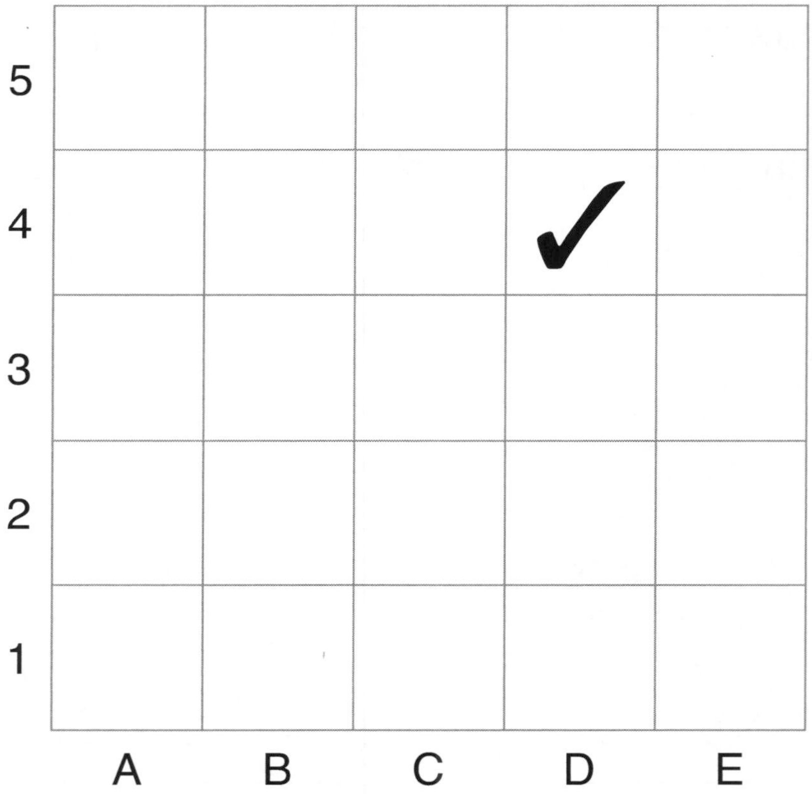

23

1 mark

24

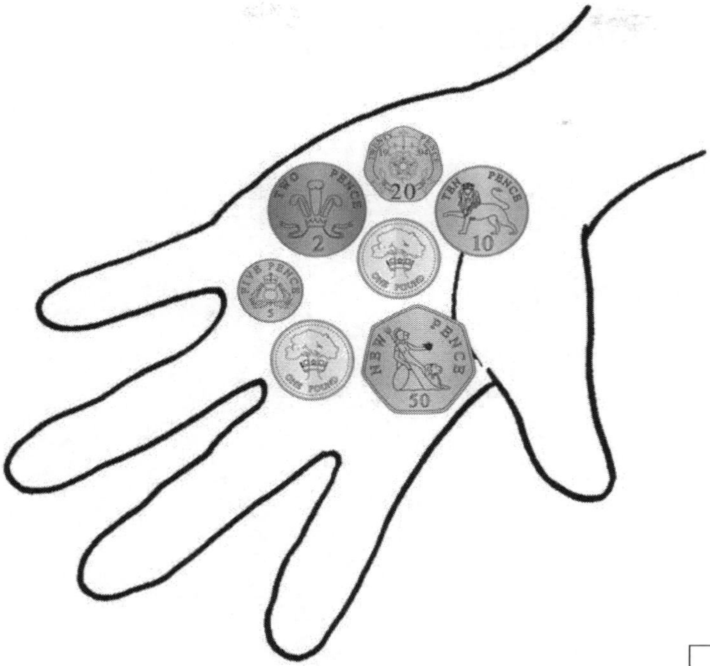

How much money is in the hand?

24

1 mark

25

Jake buys 7 packets of stickers.

There are 5 stickers in each packet.

How many stickers does he buy?

25

1 mark

26

Draw a line of symmetry through this shape.

Use your ruler.

You may use a mirror.

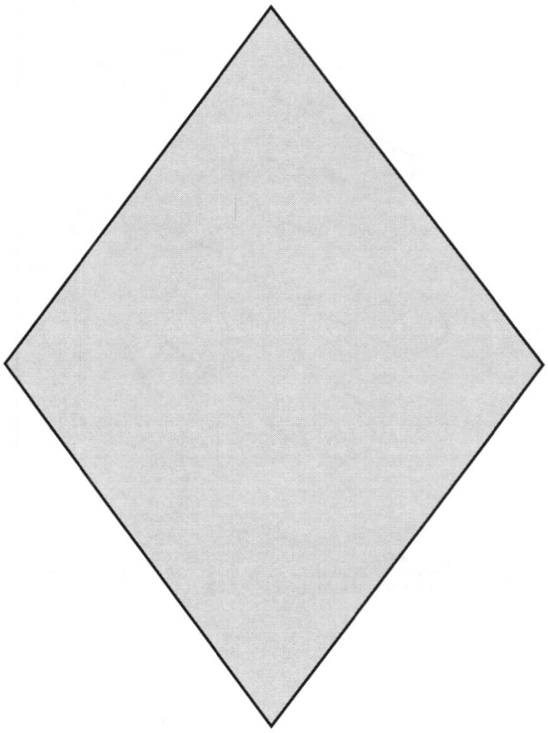

26

1 mark

27

Draw a circle around all the multiples of 5.

35	10
56	18
45	30

27

1 mark

28

Look at the bowls of fruit.

There are 18 pieces of fruit.

Circle the sum that matches the picture.

$4 + 9 + 5 = 18$

$7 + 2 + 9 = 18$

$7 + 3 + 8 = 18$

$7 + 5 + 6 = 18$

$5 + 5 + 8 = 18$

28

1 mark

29

How long is a line 3 cm **shorter** than this line?

Use a ruler.

29

1 mark

30

Look at this clock.

What time did the clock show two hours ago?

Circle it.

30

1 mark

End of test

Test 2B

Mental mathematics test questions

Say: **Listen carefully as I read aloud some questions for you to answer. I will read each question twice. The first question is a practice question which we will all do together. The boxes are for you to write your answers in. The letters below each box show you which box to use for each question.**

Question	Say
Practice	**Subtract 20 from 35.** **Write your answer in box a.**
1	**There are 43 people on a bus.** **16 get off.** **How many people are left on the bus?** **Write your answer in box b.**
2	**Jake multiplied two numbers together.** **His answer was 120.** **Which two numbers could he have multiplied together?** **Write a number in each shape to make the multiplication correct.**
3	**A cuboid has 3 green faces and 1 red face.** **All the other faces are blue.** **How many faces of the cuboid are blue?** **Write your answer in box c.**
4	**What is the sum of 38 and 43?** **Write your answer in box d.**
5	**Look at the scales.** **What is the total weight of the apples?** **Write your answer in box e.**

Year 2

Mathematics booklet

TEST 2 PAPER B | LEVEL 3

Name

Date

Class

You need
- **pencil**
- **ruler**
- **mirror**

PAGE	MARKS	PAGE	MARKS
2		10	
3		11	
4		12	
5		13	
6		14	
7		15	
8		16	
9		**TOTAL** (out of 30)	

Practice question

a

1

b

1

1 mark

2

○ × △ = 120

2

1 mark

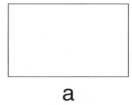

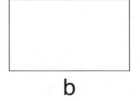

3

c

3
1 mark

4

d

4
1 mark

5

e

5
1 mark

Practice question

Write the answer.

$4 \times 5 = \square$

6

Write numbers in the boxes to make this correct.

$40 + \square = 100 = 80 + \square$

6 1 mark

7

Circle the number that is closest to 800.

870 83 751 78 850

7 1 mark

8

Freddie buys a packet of crisps.

He pays with a £2 coin.

He receives these four coins in change.

How much was the packet of crisps?

8
1 mark

9

Write the answer.

$23 \times 4 =$ ☐

9
1 mark

10 Look at the diagrams of 3-D solids.

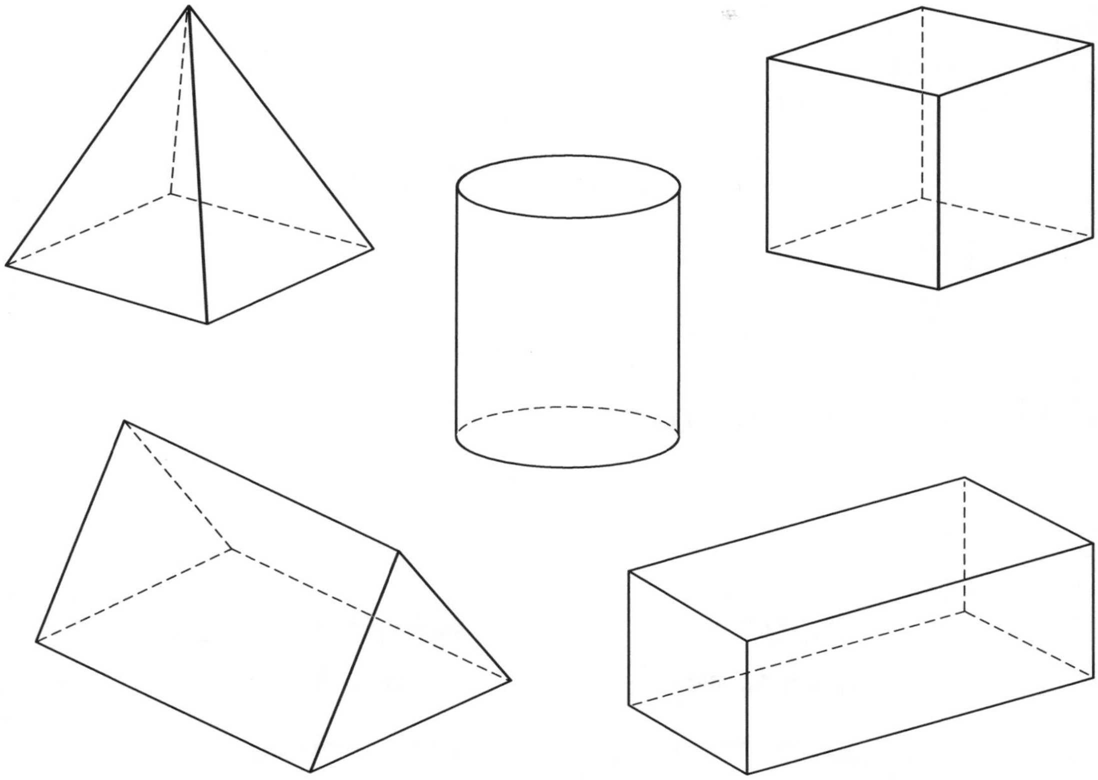

One of the shapes has three rectangular faces and two triangular faces.

Circle this shape.

10
1 mark

11 Continue the number sequence in both directions.

			23	26	29			

11
1 mark

12

Write the answer.

$$158 - 76 = \square$$

12
1 mark

13

Fill in three missing numbers.

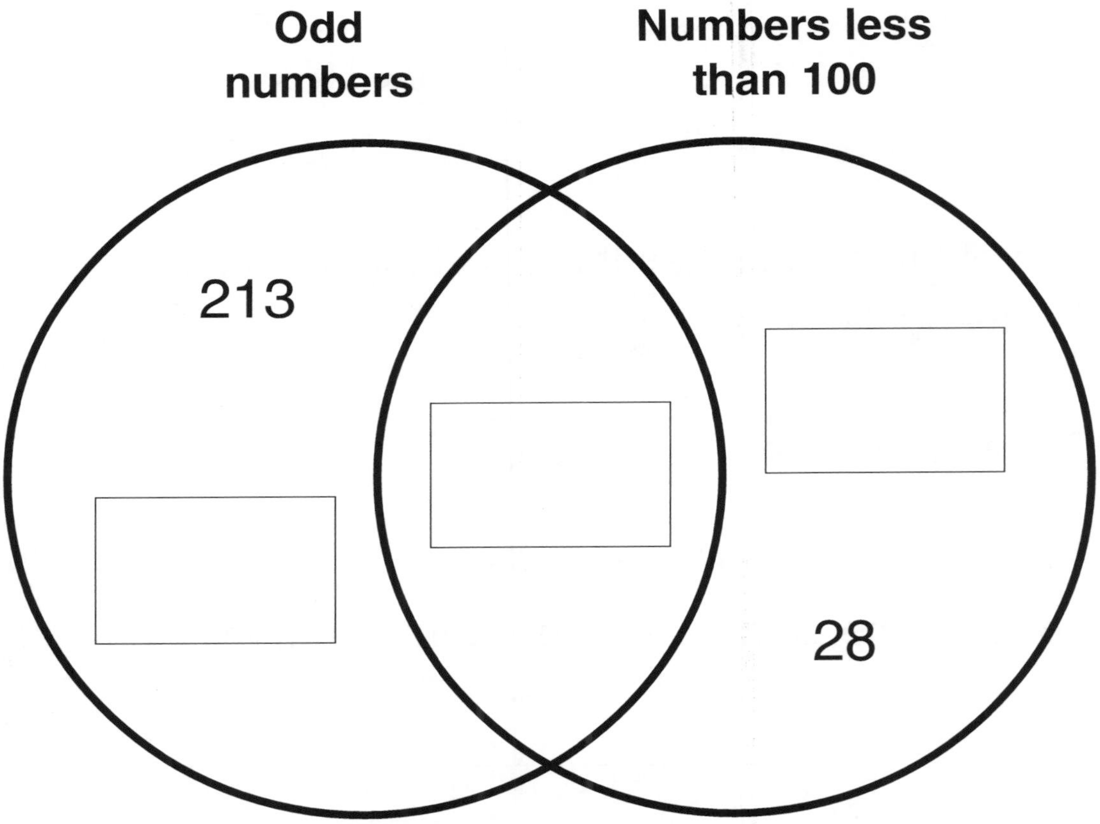

13
1 mark

14

Complete the diagram below to make a shape that is symmetrical about the mirror line.

Use a ruler.

You may use a mirror.

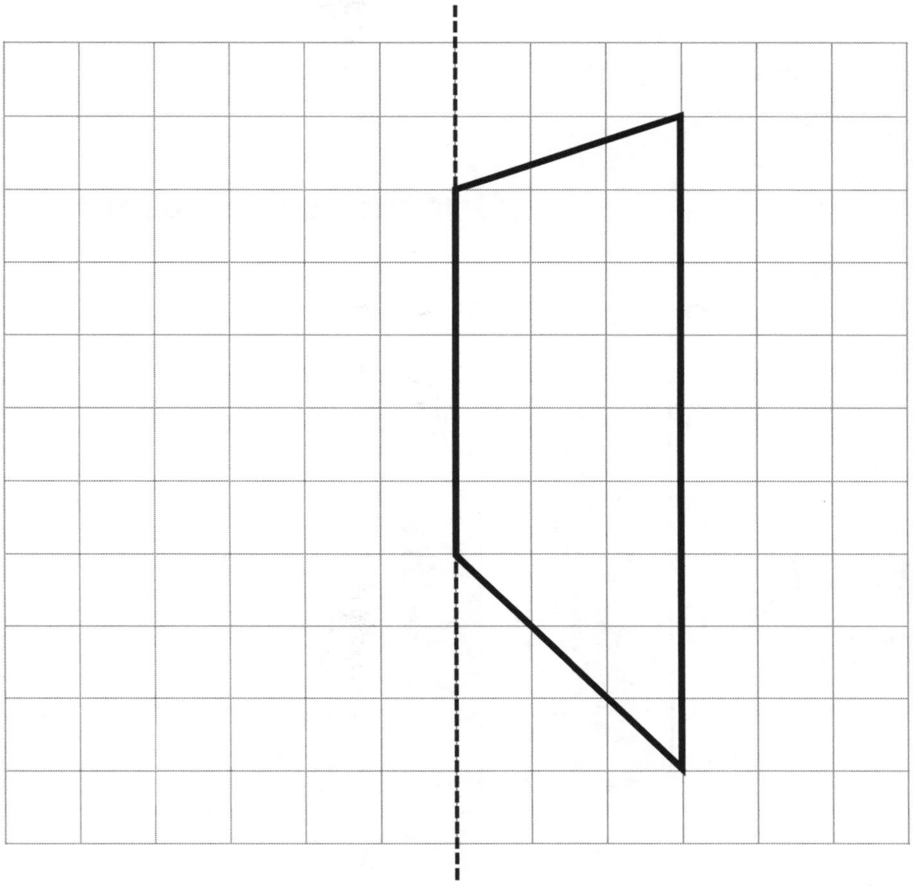

14
1 mark

15

What is half of 450?

15
1 mark

16

How many grams equal 1 kilogram?

g

16

1 mark

17

Class 2 drew a graph.

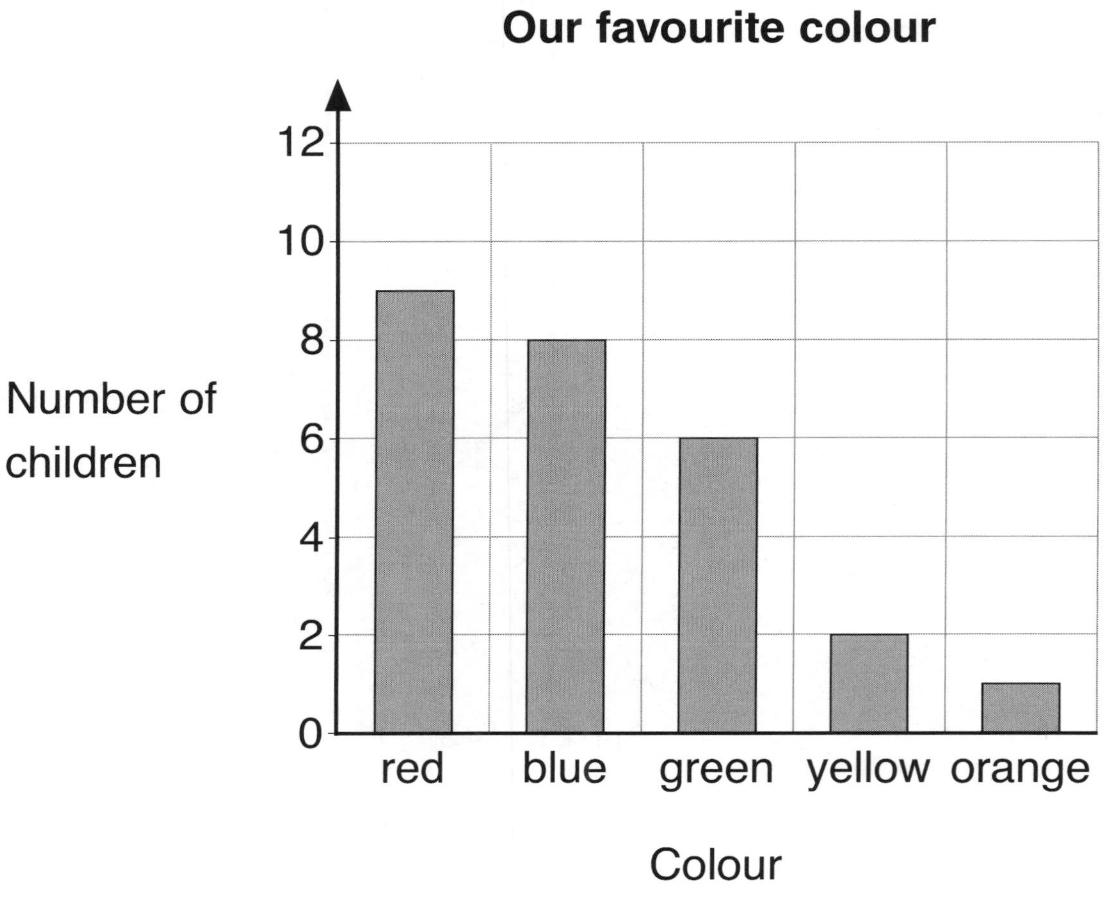

How many children are there in Class 2?

17

1 mark

18

Here is a scale which shows the weight of sugar.

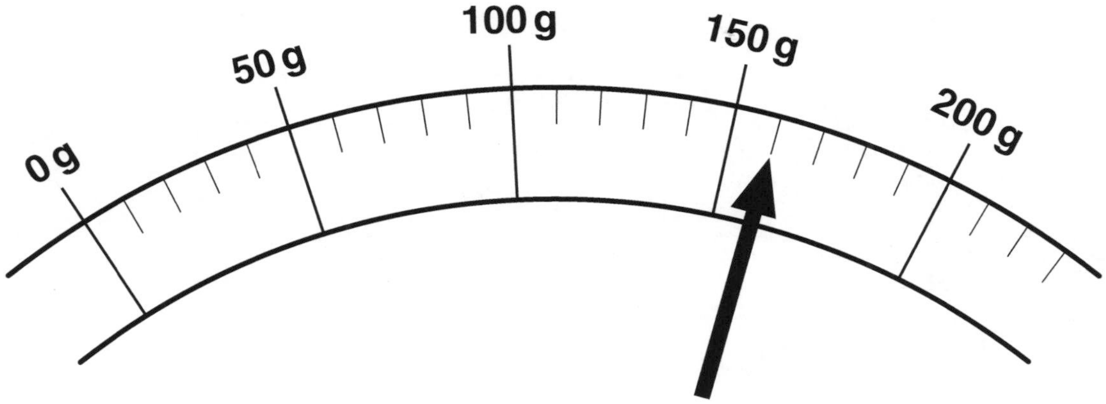

What is the weight of sugar? [g]

18
1 mark

19

Freddie needs 28 cakes for his party.

There are four cakes in a pack.

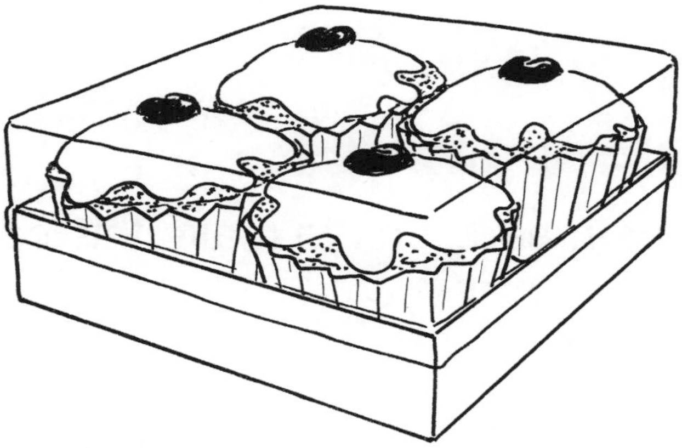

How many packs does he need to buy? [ ]

19
1 mark

20

Gita places a counter on square B3.

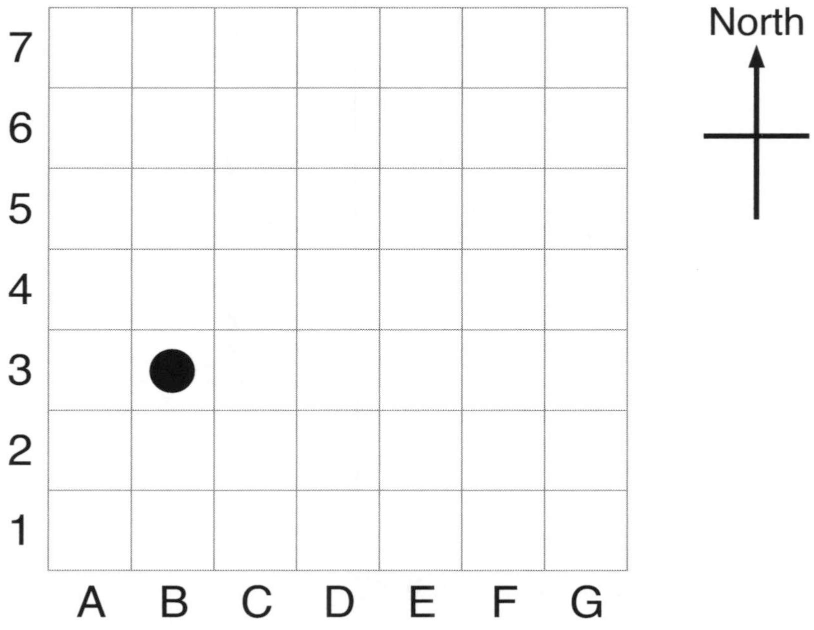

She moves it 3 squares east and 2 squares north.

Write the position of the square she moves it to.

20

1 mark

21

Write a number in each box to make this correct.

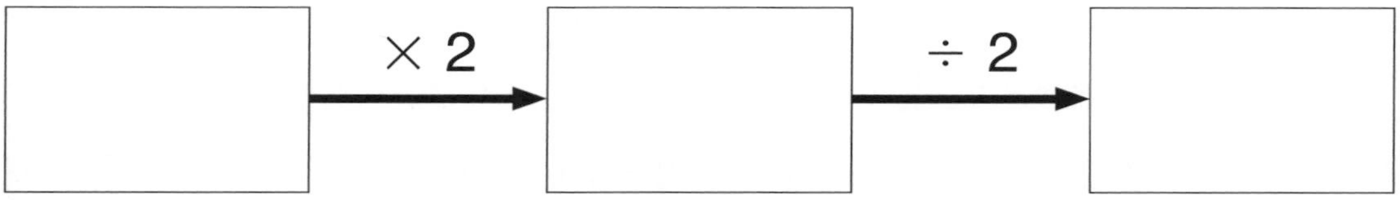

21

1 mark

22

Gita and her mum left school at this time.

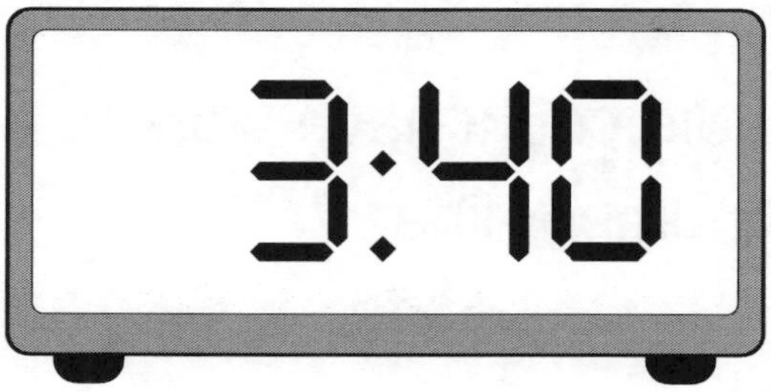

They got home 30 minutes later.

At what time did they get home?

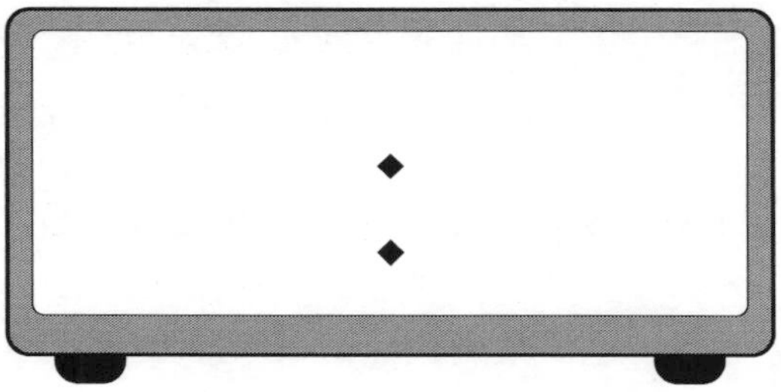

22
1 mark

23

Write the answer.

$92 - 53 = \square$

23
1 mark

24

Freddie worked out the answer to 6×4 on a number line.

Show how Freddie could have worked out the answer on this number line.

24

1 mark

25

Shade more squares so that $\frac{3}{4}$ of this shape is shaded.

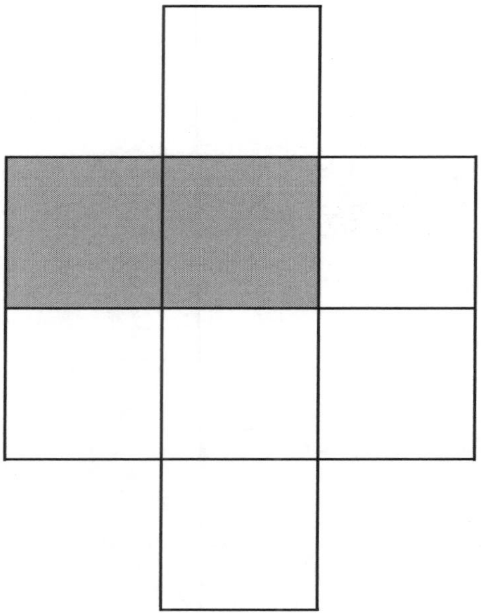

25

1 mark

Total out of 2

26

Lucy makes a sequence of numbers.

My rule is double the last number then add 2.

Write the next two numbers in her sequence.

6 14 30 [ ] [ ]

26 1 mark

27

Write the missing digits to make this correct.

[ ] 2 + 4 [ ] = 6 7

27 1 mark

28

Circle two numbers that add to make a multiple of 10.

11	12
13	14
15	16
17	18

19

28

1 mark

29

Look at these cards.

4	5	3

Use each card once to make the smallest even number.

29

1 mark

30

This is a number triangle with some numbers missing.

The numbers along each edge must add up to 90.

Put all the numbers 10, 30, 40 and 60 in the circles to make the totals correct.

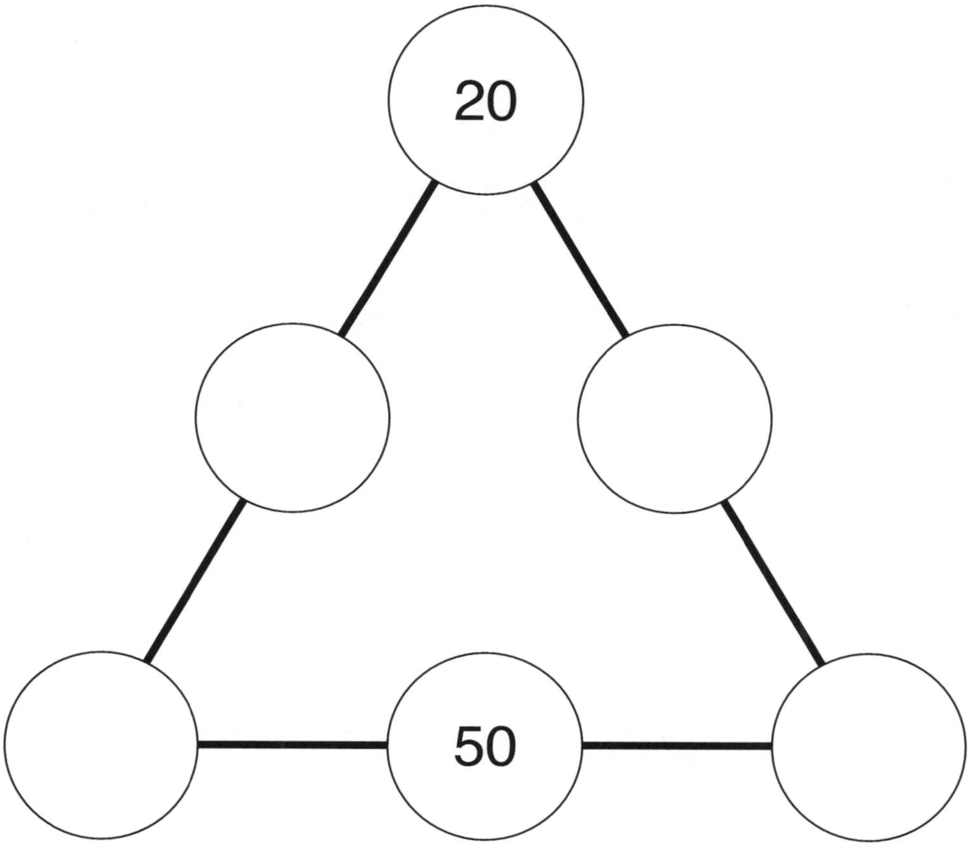

30

1 mark

End of test

Test 3

Papers A and B answers

Paper A

1. 23
2. 30
3. 
4. 15 ÷ 3
5. 2 litres
6. *Accept* 41, 42, 43, *or* 45
7. 61, 56, 51, **46**, **41**, 36, **31**, 26
8. 35 litres
9. *Check that the child has shaded 4 more squares.*
10. 2 + 6 + 7 = 15
11. 80
12. 5
13. 18
14. 7 litres
15. 7
16. right 5, down 5, left 8
17. 8 corners, 6 faces, 12 edges
18. 17
19. £2.49
20. 20
21. 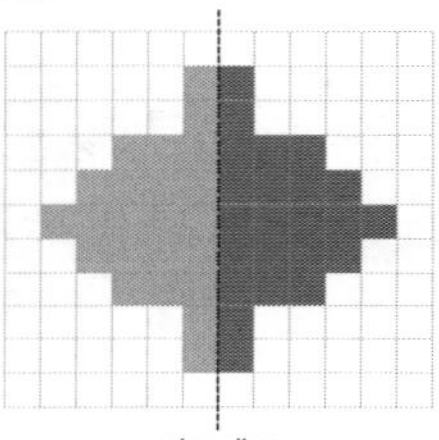

22. 8 − 3 = 5
23. **29**, 31, 33, 35, 37, **39**, 41, 43
24. 2 × 5 = 10, 5 × 2 = 10, 10 ÷ 2 = 5, 10 ÷ 5 = 2
25. 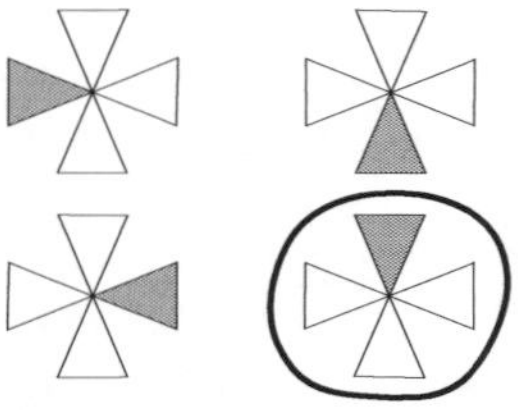
26. 26
27. 
28. 5
29. 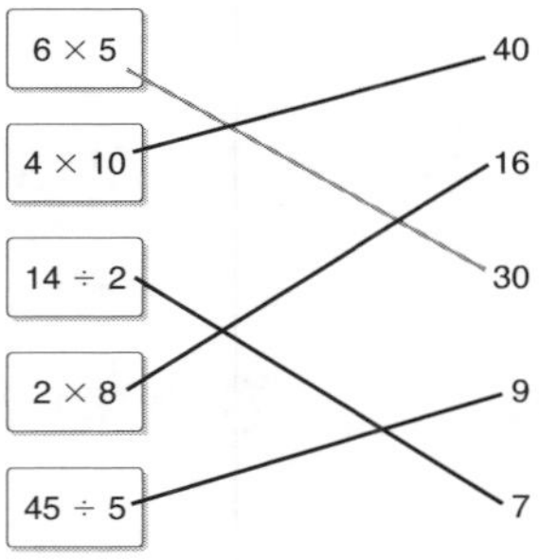

30. 50

Paper B

1. £45
2. 451
3. 450 ml
4. 12
5. 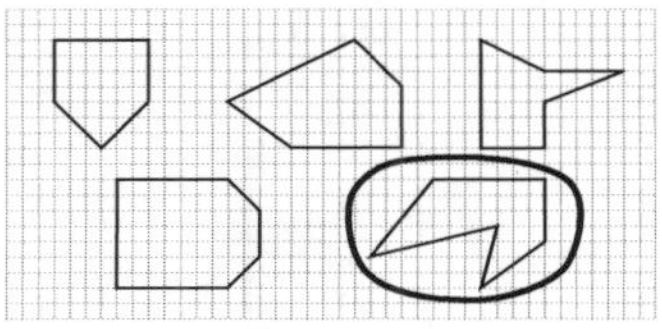
6. 136, 341, 438, 504, 552
7. *Accept calculation where any two numbers added to 460 total 530, e.g.* 460 + **50** + **20** = 530
8. 3
9. *Accept any of the following:*
   40 ÷ **20** = **2**
   40 ÷ **10** = **4**
   40 ÷ **8** = **5**
   40 ÷ **5** = **8**
   40 ÷ **4** = **10**
   40 ÷ **2** = **20**
10.

	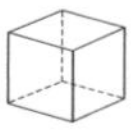 cube	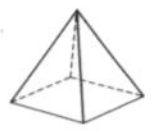 square-based pyramid
number of faces	6	5
number of edges	12	8
number of corners	8	5

11. 396
12. 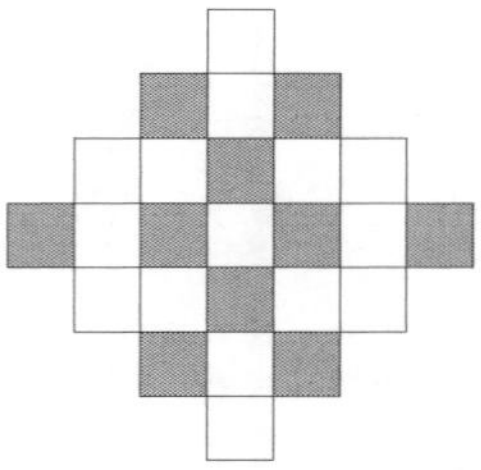
13. 838
14. 40
15. 29
16. 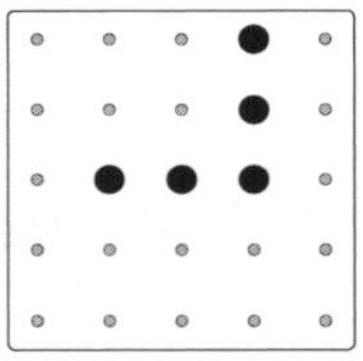
17. *Check that the child has shaded any 6 triangles.*
18. 5, 11, 17, **23**, 29, 35
19. C and E
20. 1 m 20 cm
21. *Check child's method, e.g.*
    80 = 40 + 40
    40 ÷ 5 = 8
    40 ÷ 5 = 8
    8 + 8 = 16
22. 72
23. 150 ml
24. 
25. 63
26. 9
27. 331
28. £4.75
29. 16
30. add 9

Test 3A

Mental mathematics test questions

Say: **Listen carefully as I read aloud some questions for you to answer. I will read each question twice. The first question is a practice question which we will all do together. The boxes are for you to write your answers in. The letters below each box show you which box to use for each question.**

Question	Say
Practice	**Look at box a.** **Write the number 25 in box a.**
1	**Listen to this sequence: 11, 14, 17, 20,** [one clap] **26** **Write the missing number in box b.**
2	**There are 5 tables in the classroom.** **6 children sit at each table.** **How many children are there in the classroom?** **Write your answer in box c.**
3	**Look at the shapes.** **Two of these shapes are not hexagons.** **Circle each shape which is not a hexagon.**
4	**Look at the number sentences.** **Circle the number sentence that has an answer of 5.**
5	**Look at the bottle of lemonade.** **Circle the label that shows how much the bottle holds.**

Year 2

Mathematics booklet

TEST 3 PAPER A | **LEVEL 2**

Name

Date

Class

You need
- **pencil**
- **ruler**
- **mirror**

PAGE	MARKS	PAGE	MARKS
2		10	
3		11	
4		12	
5		13	
6		14	
7		15	
8		16	
9		**TOTAL** (out of 30)	

Practice question

a

1

b

1

1 mark

2

c

2

1 mark

3

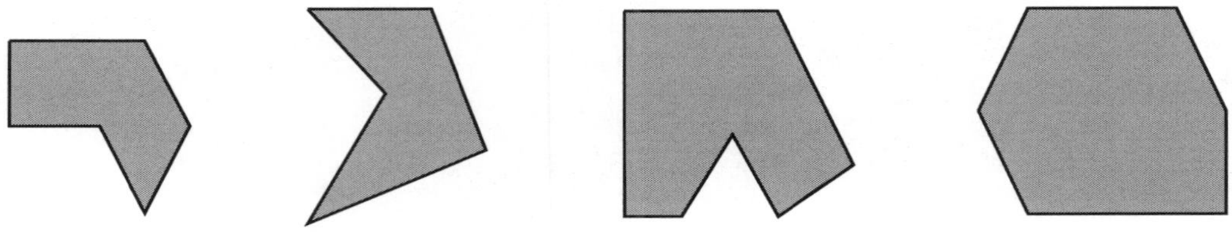

3

1 mark

4

14 − 8

15 ÷ 3

2 × 3

3 + 4

4

1 mark

5

2 grams

2 metres

2 kilograms

2 litres

2 centimetres

5

1 mark

Total out of 3

Practice question

Write the answer.

What is 10 more than 25? [ ]

6 Look at these cards.

3 4 2 5 1

Use two of the cards to make a number between 40 and 50.

[ ] [ ]

6 1 mark

7 Write the missing numbers in the sequence.

61	56	51			36		26

7 1 mark

8

Kevin's dad washes the windows of their home.

He uses 7 buckets of water.

Each bucket has 5 litres of water.

How many litres of water does he use altogether?

litres

8

1 mark

9

Three of the squares in this shape are shaded.

Shade more squares so that exactly half of the shape is shaded.

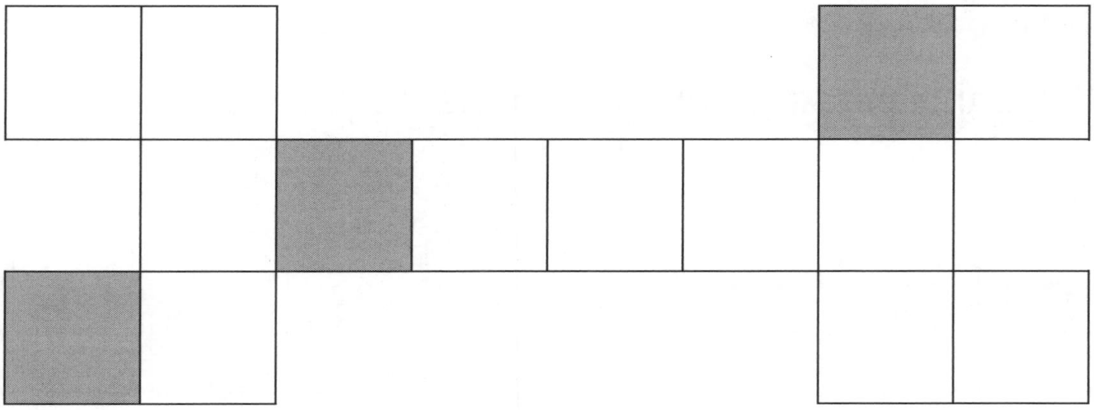

9

1 mark

Total out of 2

10

Look at the number line.

It shows the sum that Jane did.

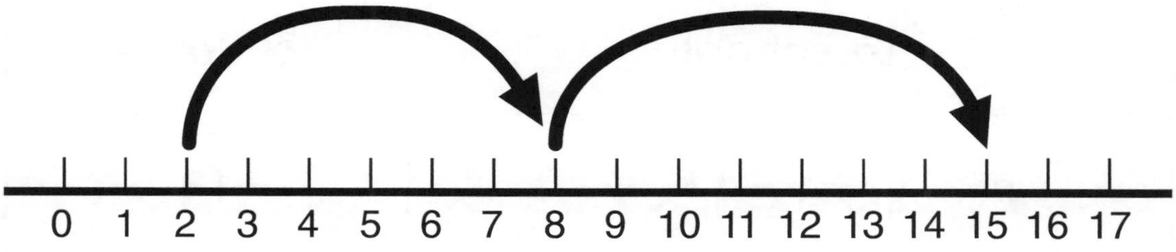

Circle the sum that Jane did.

$2 + 5 + 8 = 15$

$2 + 8 + 5 = 15$

$2 + 6 + 7 = 15$

$2 + 9 + 4 = 15$

10
1 mark

11

Write the correct number in the box. One has been done for you.

42	to the nearest 10 is	40
78	to the nearest 10 is	

11
1 mark

12

This table shows how the children in a class come to school.

Ways of coming to school		Number of children
walk		13
taxi		3
bus		6
car		8

More children walk to school than come by car.

How many more? ☐

12
1 mark

13

Write a number in the box to make this correct.

$$60 + \square = 78$$

13
1 mark

14

How much water is in the bucket? litres

14
1 mark

15

Write a number in the box to make this correct.

$9 + 4 = 6 + \square$

15
1 mark

16

Jane slid her finger along this route from START to STOP.

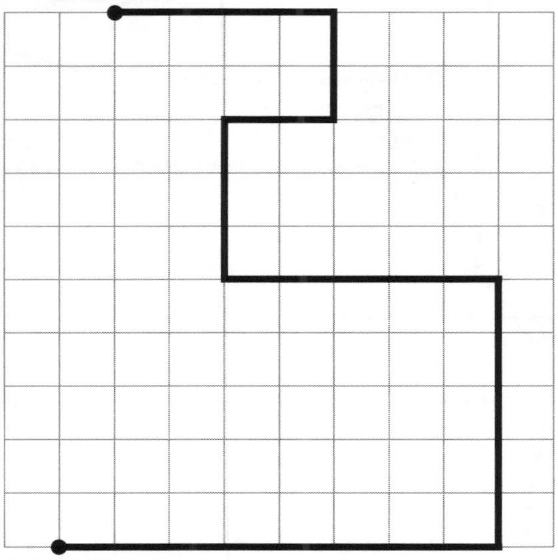

She started writing how her finger moved.

Complete the moves.

right 4

down 2

left 2

down 3

____

____

____

16

1 mark

17

Write each word in the correct box.

faces	edges	corners

A cube has

- 8 ☐
- 6 ☐
- 12 ☐

17

1 mark

18

Write the total.

$7 + 2 + 3 + 5 = \square$

18

1 mark

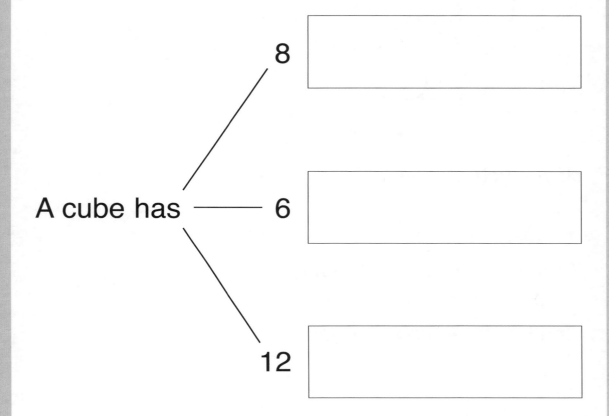

19

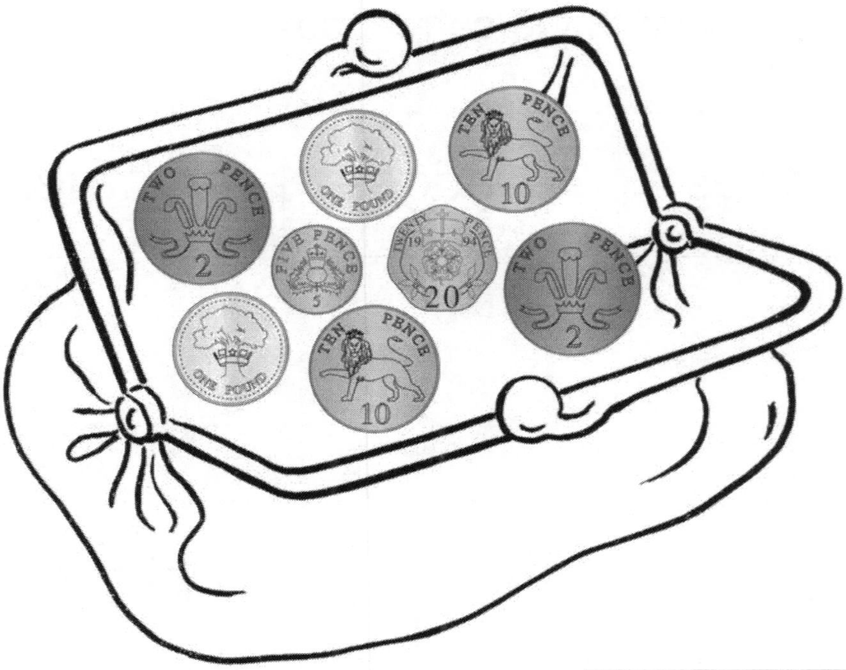

How much money is in the purse?

19

1 mark

20

Write in the missing number.

One has been done for you.

4 ➔ double and add 5 ➔ 13

7 ➔ double and add 6 ➔ ☐

20

1 mark

21

Show the reflection of this pattern in the mirror line.

You can use a mirror.

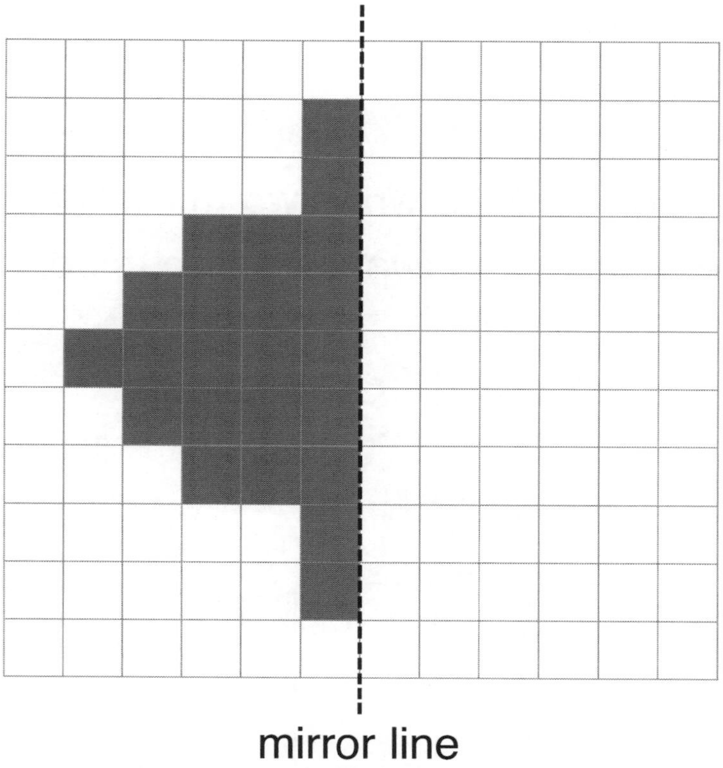

21

1 mark

22

Only one of these is correct.

Circle it.

$6 + 5 = 10$

$12 - 7 = 6$

$10 + 9 = 18$

$8 - 3 = 5$

22

1 mark

23

Fill in the two missing numbers in this sequence.

[ ] 31 33 35 37 [ ] 41 43

23
1 mark

24

Use these digit cards to write two multiplication number sentences and two division number sentences.

2 5 10

[ ] × [ ] = [ ]

[ ] × [ ] = [ ]

[ ] ÷ [ ] = [ ]

[ ] ÷ [ ] = [ ]

24
1 mark

25 What will this shape look like after a half turn?

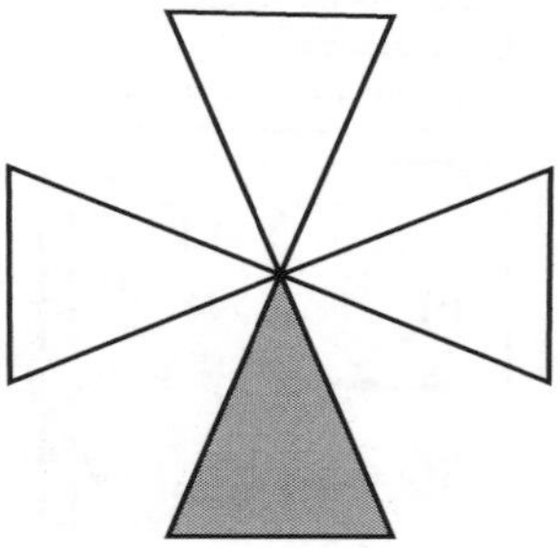

Circle the shape which shows this.

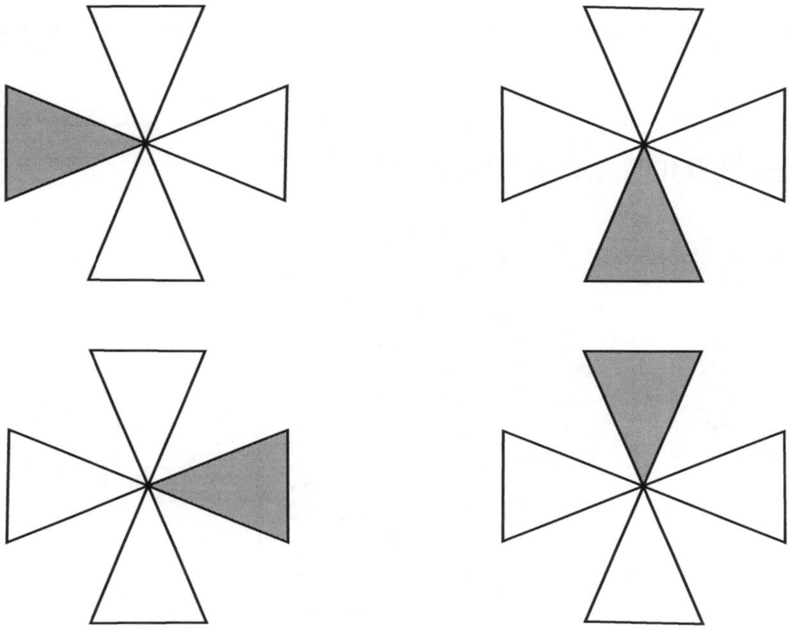

25

1 mark

26 Write the answer.

63 − 37 = ☐

26

1 mark

27

Kevin's watch shows this time.

What time will his watch show three hours later?

Draw the hands on the watch.

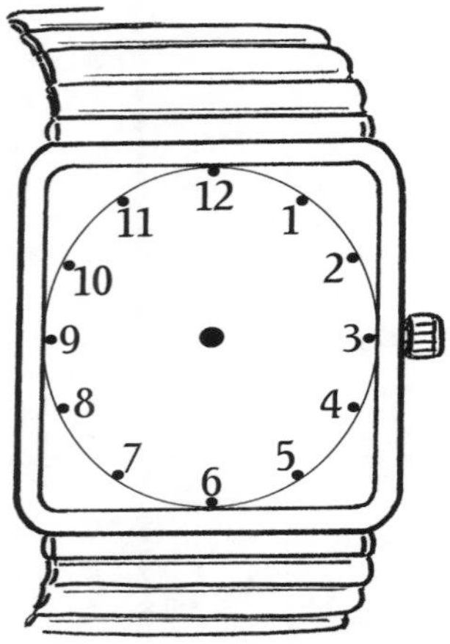

27
1 mark

28

20 children sit at tables in groups of 4.

How many groups are there?

28
1 mark

Total out of 2

29 Match each card to an answer.

One has been done for you.

6×5	40
4×10	16
$14 \div 2$	30
2×8	9
$45 \div 5$	7

29
1 mark

30 Write the missing number in the box.

$\square \div 5 = 10$

30
1 mark

End of test

Test 3B

Mental mathematics test questions

Say: **Listen carefully as I read aloud some questions for you to answer. I will read each question twice. The first question is a practice question which we will all do together. The boxes are for you to write your answers in. The letters below each box show you which box to use for each question.**

Question	Say
Practice	**Add 40 and 30.** **Write your answer in box a.**
1	**A jumper costs £27.** **A pair of jeans costs £18.** **What is the total cost of a jumper and a pair of jeans?** **Write your answer in box b.**
2	**Look at the numbers in box c.** **Circle the number that is closest to 500.**
3	**Look at the jug of water.** **How many more millilitres do you need to add to the jug to make 1 litre?** **Write your answer in box d.**
4	**Three is a quarter of a number.** **What is the number?** **Write the number in box e.**
5	**Look at the shapes.** **One of these shapes is a hexagon that has one right angle.** **Circle the correct shape.**

Year 2

Mathematics booklet

TEST 3 PAPER B | LEVEL 3

Name

Date

Class

You need
- **pencil**
- **ruler**

PAGE	MARKS	PAGE	MARKS
2		10	
3		11	
4		12	
5		13	
6		14	
7		15	
8		16	
9		**TOTAL** (out of 30)	

Practice question

a

1

£

b

1

1 mark

2

570 52 451 49 550

c

2

1 mark

3

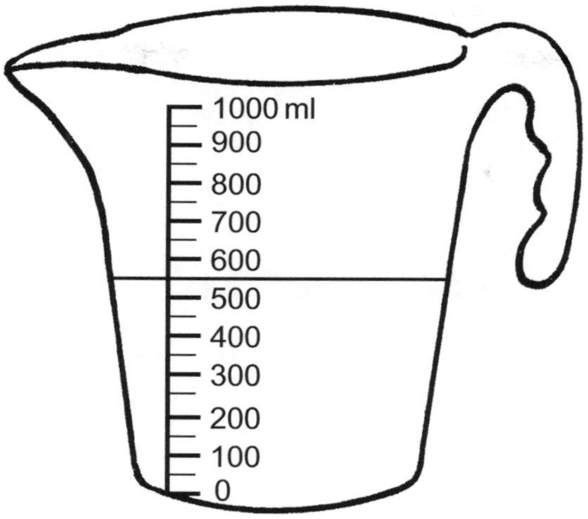

ml

d

3
1 mark

4

e

4
1 mark

5

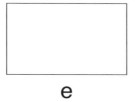

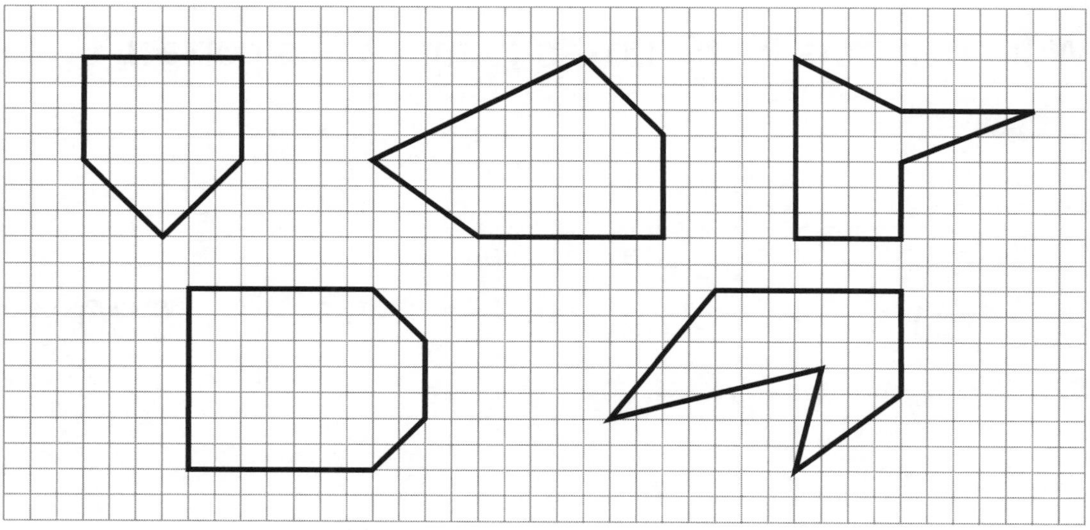

5
1 mark

Practice question

Write the answer.

What is 10 more than 25? [ ]

6 Write these numbers in order of size.

341 552 438 504 136

smallest ... largest

[ ] [ ] [ ] [ ] [ ]

6 1 mark

7 Write numbers in the boxes to make this correct.

$460 + \square + \square = 530$

7 1 mark

8

A group of children were asked how many brothers and sisters they have.

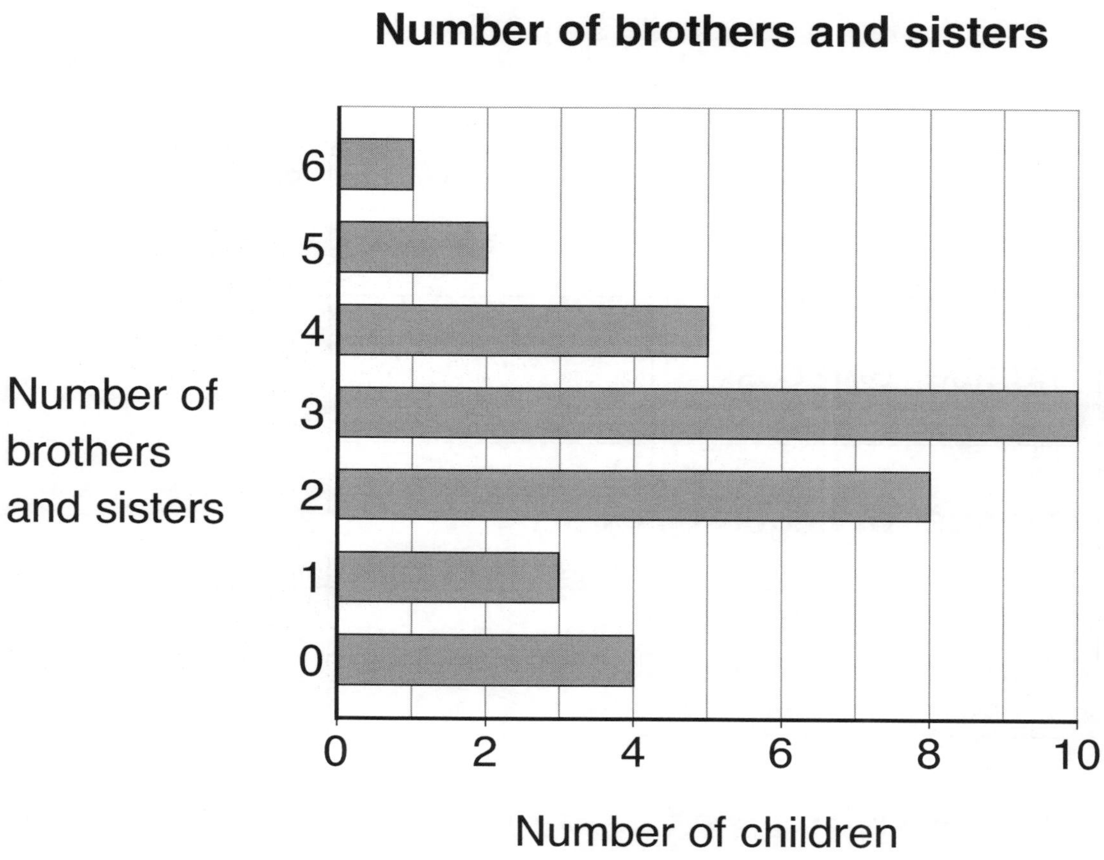

How many children have more than 4 brothers and sisters?

8
1 mark

9

Write numbers in the boxes to make this correct.

$$40 \div \square = \square$$

9
1 mark

10

Complete the table.

	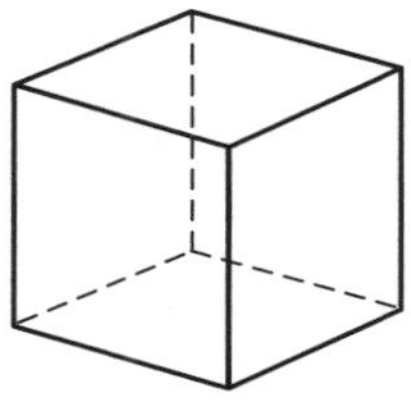 cube	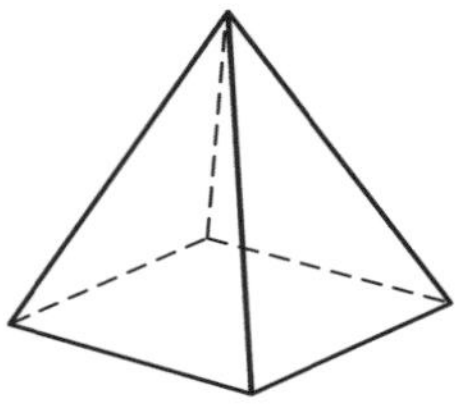 square-based pyramid
number of faces		
number of edges		
number of corners		

10

1 mark

11

Here are some numbers.

456 347 412 396 408

Write one of the numbers in the box to make this correct.

The number ☐ rounded to the nearest 10 is 400.

11

1 mark

12

Here is a grid with eight squares shaded.

Shade two more squares to make a symmetrical pattern.

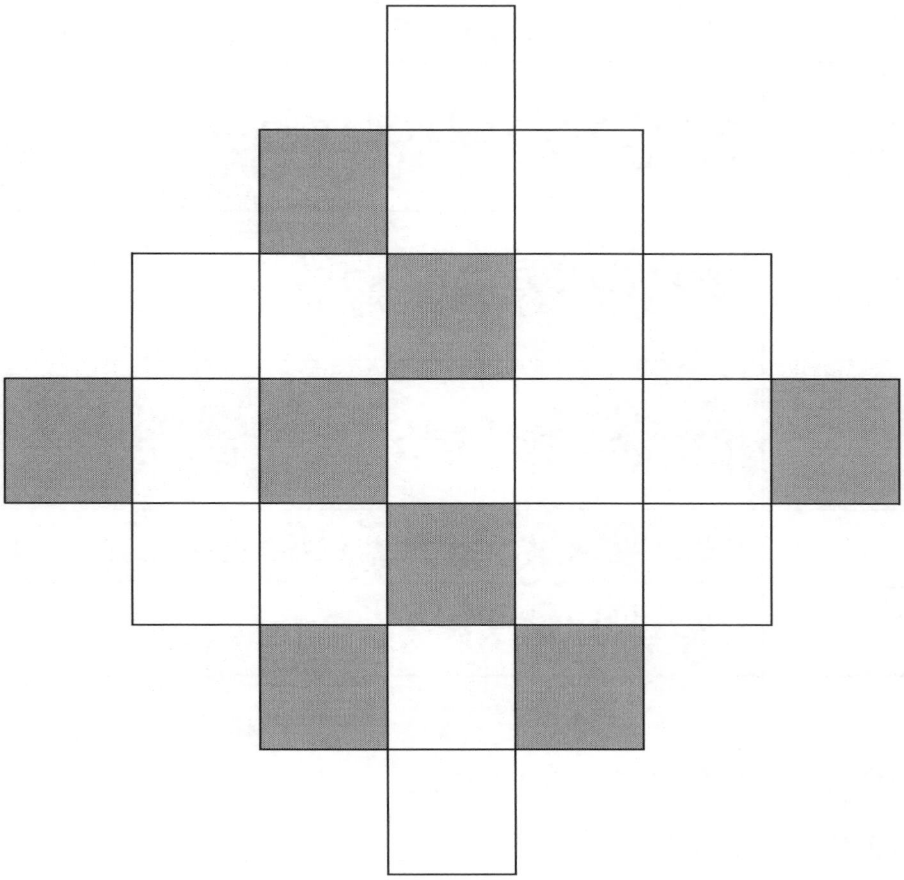

12
1 mark

13

Write the answer.

$1000 - 162 = \square$

13
1 mark

14

Some children drew a pictogram after a visit to the local park.

Flowers we saw

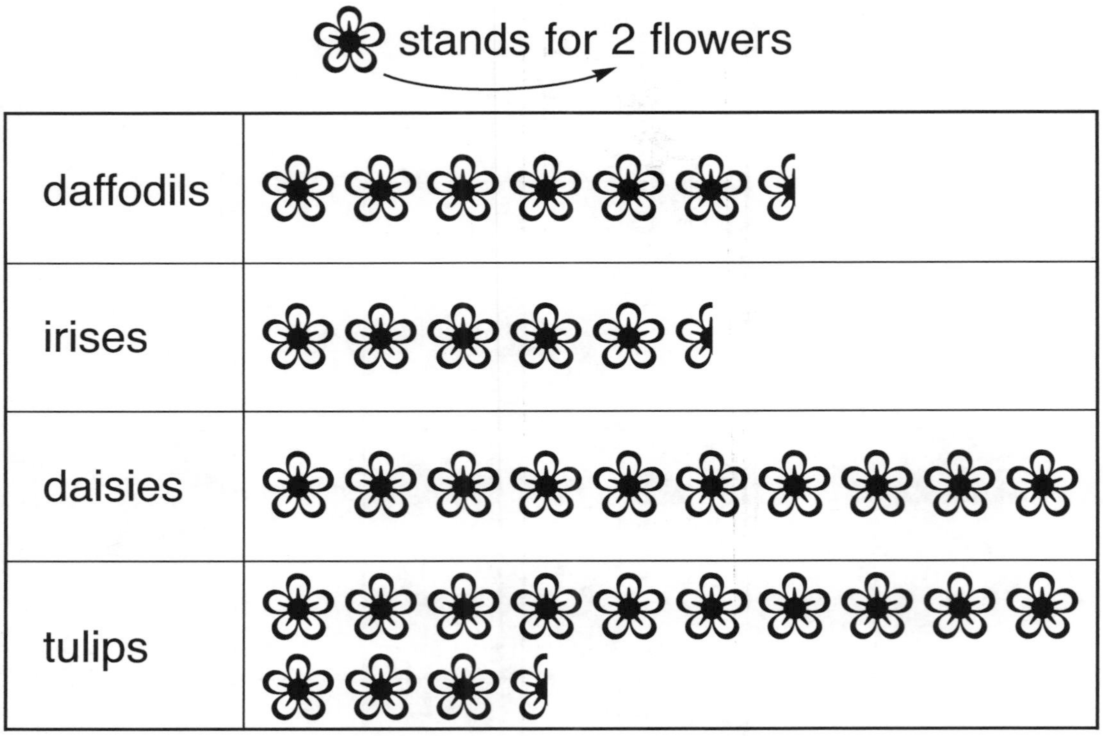

How many daffodils and tulips did the children see?

14 1 mark

15

Work out the number halfway between 18 and 40.

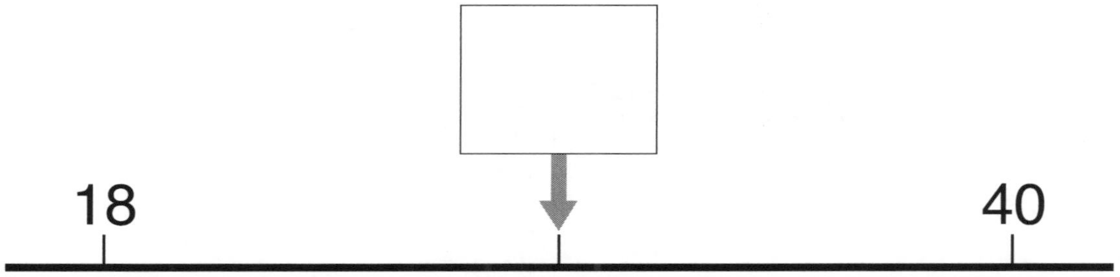

15 1 mark

16

Josh put five pegs in a pegboard.

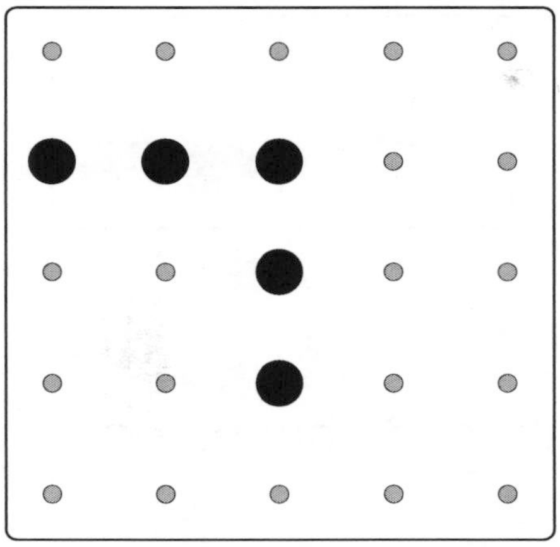

He turns the board clockwise through 1 right angle.

Draw how the board looks now.

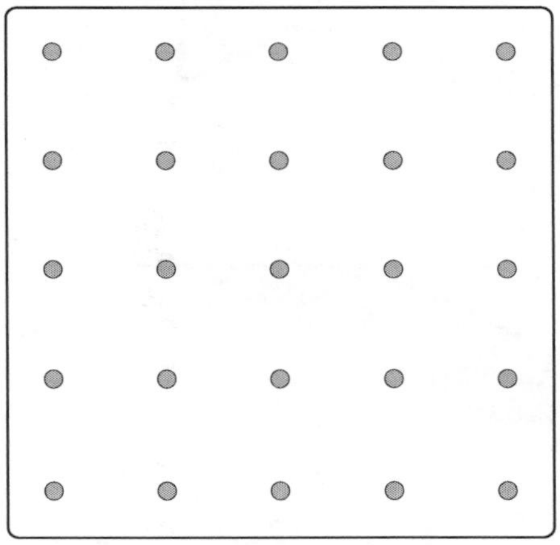

16

1 mark

17

Shade $\frac{3}{4}$ of this shape.

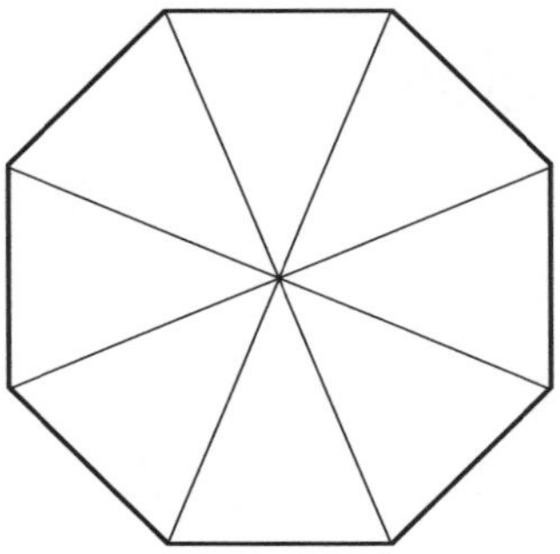

17

1 mark

Total out of 2

18

Write the missing number in this sequence.

5	11	17		29	35

18

1 mark

19

Here are some shaded shapes on a square grid.

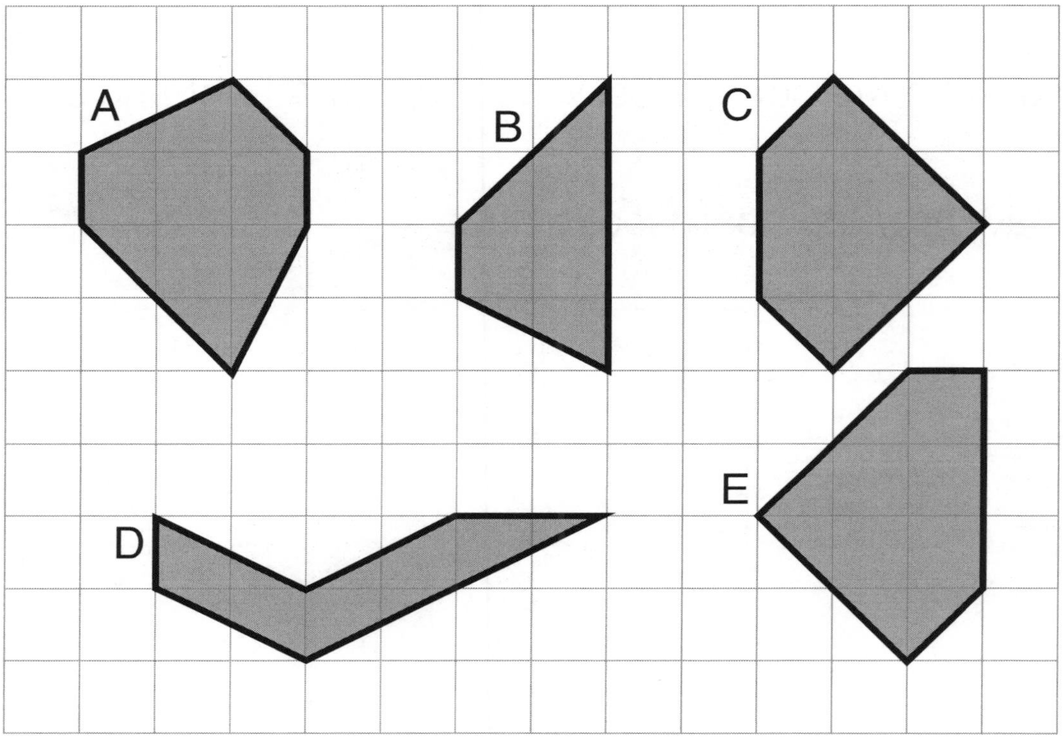

Write the letters of the two shapes which have right angles.

[ ] and [ ]

19

1 mark

20

Josh has 4 m 80 cm of rope.

He cuts it into four equal pieces.

How long is each piece? ☐

20
1 mark

21

Ellie worked out the correct answer to **80 ÷ 5**

Her answer was **16**.

Show how she could have worked out her answer.

16

21
1 mark

22

Write the answer.

$24 \times 3 =$ ☐

22

1 mark

23

This jug has water in it.

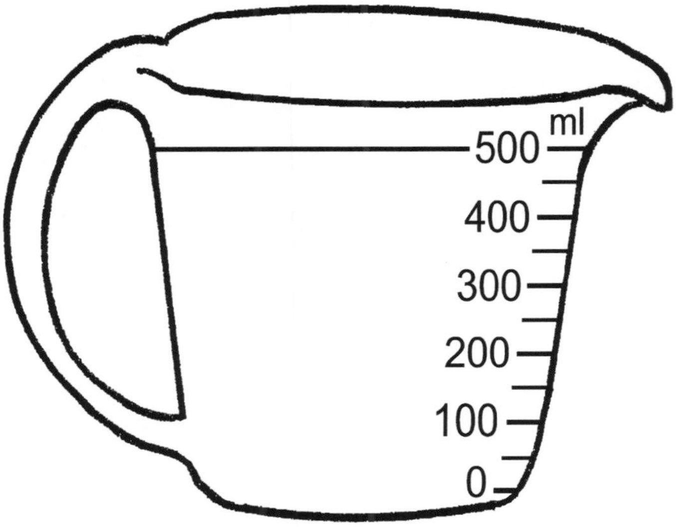

Ellie pours 350 ml of water out of this jug.

How much water will be left in the jug?

☐ ml

23

1 mark

24

The cake started to cook at this time.

It cooked for 50 minutes.

At what time did it finish cooking?

24

1 mark

25

Write the missing number in the box.

$\square + 37 = 100$

25

1 mark

26

A carton of squash fills 6 cups.

Mrs White wants to fill 50 cups of squash.

How many cartons of squash does she need to buy?

26

1 mark

27

Write the total.

$$256 + 75 = \square$$

27

1 mark

28

Ellie and Josh go to the cinema.

They buy one popcorn and two drinks.

popcorn
£2.35

drink
£1.20

How much do they spend altogether?

Show how you work it out.

28

1 mark

29

Ellie ate half the biscuits on the plate.

These are the biscuits that were left.

How many biscuits were on Ellie's plate before she ate half of them?

29

1 mark

30

Look at this sequence.

117 126 135 144 153

Circle the rule for this sequence.

subtract 8	add 12
add 9	divide by 2

30

1 mark

End of test

Record-keeping format 1 Adult Directed Task assessment sheet

Objective(s): ______________________ Date: __________ Adult: __________

______________________ NC Level: __________ Class: __________

Child's name	Success criteria				Other observations	Objective(s) achieved	Future action

Record-keeping format 2 Test 1 grid for test analysis (Paper A: Level 2)

	Number sequence	Addition	Properties of 2-D shapes	Multiplication	Calculating time differences	Multiplication	Rounding	Identifying which operation to use	Halving: money	Reflective symmetry	Ordering whole numbers	Addition of more than two numbers	Handling data: table	Addition
AT	2	2	3	1 & 2	3	2	2	2	1 & 2	3	2	2	2	2
Question	1	2	3	4	5	6	7	8	9	10	11	12	13	14
Mark	1	1	1	1	1	1	1	1	1	1	1	1	1	1
Name														
1.														
2.														
3.														
4.														
5.														
6.														
7.														
8.														
9.														
10.														
11.														
12.														
13.														
14.														
15.														
16.														
17.														
18.														
19.														
20.														
21.														
22.														
23.														
24.														
25.														
26.														
27.														
28.														
29.														
30.														
Number correct														
Number incorrect or omitted														
Percentage correct														
Percentage incorrect or omitted														

Record-keeping format 2 Test 1 grid for test analysis (Paper A: Level 2)

	Comparing lengths	Properties of numbers: odd and even	Division	Calculating time intervals	Identifying numbers on a number line	Handling data: pictogram	Rotating patterns	Understanding relationship between + and −	Number sequences	Naming 2-D shapes and 3-D solids	Finding half of a shape	Subtraction (difference)	Drawing lines to the nearest cm	Doubling	Identifying multiples of 10	Handling data: tally charts	Paper A score (out of 30)	**National Curriculum Level**
	3	2	1 & 2	1 & 3	2	2	3	2	2	3	2	2	3	2	2	2		
	15	16	17	18	19	20	21	22	23	24	25	26	27	28	29	30		
	1	1	1	1	1	1	1	1	1	2	1	1	1	1	1	1		
1.																		
2.																		
3.																		
4.																		
5.																		
6.																		
7.																		
8.																		
9.																		
10.																		
11.																		
12.																		
13.																		
14.																		
15.																		
16.																		
17.																		
18.																		
19.																		
20.																		
21.																		
22.																		
23.																		
24.																		
25.																		
26.																		
27.																		
28.																		
29.																		
30.																		

Record-keeping format 2 Test 1 grid for test analysis (Paper B: Level 3)

	Subtraction	Addition	Recognising right angles	Writing 3-digit numbers	Reading scales: length	Understanding addition and subtraction	Properties of numbers: 3-digit even numbers	Addition: money	Halving	Division	Estimating a number on a number line	Addition of more than two numbers	Reading time: analogue and digital	Identifying squares on a grid
AT	2	1 & 2	3	2	3	2	2	1 & 2	2	1 & 2	2	2	3	3
Question	1	2	3	4	5	6	7	8	9	10	11	12	13	14
Mark	1	1	1	1	1	1	1	1	1	1	1	1	1	1
Name														
1.														
2.														
3.														
4.														
5.														
6.														
7.														
8.														
9.														
10.														
11.														
12.														
13.														
14.														
15.														
16.														
17.														
18.														
19.														
20.														
21.														
22.														
23.														
24.														
25.														
26.														
27.														
28.														
29.														
30.														
Number correct														
Number incorrect or omitted														
Percentage correct														
Percentage incorrect or omitted														

Record-keeping format 2 Test 1 grid for test analysis (Paper B: Level 3)

	Subtraction: length	Understanding repeated addition as multiplication	Place value: 10 more than	Number sequences: money	Addition and subtraction	Identifying shapes with and without symmetry	Ordering whole numbers	Addition of more than two numbers	Explaining calculation methods	Identifying 2-D shapes	Subtraction	Handling data: Carroll diagram	Finding half of a shape	Length: measuring and drawing lines	Addition	Handling data: tally chart	Paper B score (out of 30)	**National Curriculum Level**
	1, 2 & 3	2	2	2	1 & 2	3	2	2	2	3	2	1 & 2	2	3	2	2		
	15	16	17	18	19	20	21	22	23	24	25	26	27	28	29	30		
	1	1	1	1	1	1	1	1	1	1	1	1	1	1	1	1		
1.																		
2.																		
3.																		
4.																		
5.																		
6.																		
7.																		
8.																		
9.																		
10.																		
11.																		
12.																		
13.																		
14.																		
15.																		
16.																		
17.																		
18.																		
19.																		
20.																		
21.																		
22.																		
23.																		
24.																		
25.																		
26.																		
27.																		
28.																		
29.																		
30.																		

Record-keeping format 3 Test 2 grid for test analysis (Paper A: Level 2)

Name	Repeated addition	Addition	Recognising 3-D solids	Recognising 2-digit numbers	Reading scales: weight	Properties of number	Using < and > signs	2-D shapes	Rounding to the nearest 10	Place value / addition	Number sequence	Handling data: bar chart	Adding more than two multiples of 10	Making 3-digit numbers
AT	2	1 & 2	3	2	3	2	2	3	2	1 & 2	2	2	2	1 & 2
Question	1	2	3	4	5	6	7	8	9	10	11	12	13	14
Mark	1	1	1	1	1	1	1	1	1	1	1	1	1	1
1.														
2.														
3.														
4.														
5.														
6.														
7.														
8.														
9.														
10.														
11.														
12.														
13.														
14.														
15.														
16.														
17.														
18.														
19.														
20.														
21.														
22.														
23.														
24.														
25.														
26.														
27.														
28.														
29.														
30.														
Number correct														
Number incorrect or omitted														
Percentage correct														
Percentage incorrect or omitted														

Record-keeping format 3 Test 2 grid for test analysis (Paper A: Level 2)

	Addition	Reading scales: weight	Identifying the unknown number in a number sentence	Handling data: pictogram	Multiplication	Subtraction	Rotation of shape	Finding one quarter of a number	Identifying squares on a grid	Addition: money	Multiplication	Identifying lines of symmetry on a shape	Identifying multiples of 5	Identifying the correct number sentence	Length: measuring lines to the nearest cm / subtraction	Calculating time intervals
	2	3	2	2	2	1 & 2	3	1 & 2	3	1 & 2	1 & 2	3	2	1 & 2	3	3
	15	16	17	18	19	20	21	22	23	24	25	26	27	28	29	30
	1	1	1	1	1	1	1	1	1	1	1	1	1	1	1	1
1.																
2.																
3.																
4.																
5.																
6.																
7.																
8.																
9.																
10.																
11.																
12.																
13.																
14.																
15.																
16.																
17.																
18.																
19.																
20.																
21.																
22.																
23.																
24.																
25.																
26.																
27.																
28.																
29.																
30.																

	Paper A score (out of 30)	**National Curriculum Level**
1.		
2.		
3.		
4.		
5.		
6.		
7.		
8.		
9.		
10.		
11.		
12.		
13.		
14.		
15.		
16.		
17.		
18.		
19.		
20.		
21.		
22.		
23.		
24.		
25.		
26.		
27.		
28.		
29.		
30.		

Record-keeping format 3 Test 2 grid for test analysis (Paper B: Level 3)

	Subtraction	Multiplication (inverse)	Properties of 3-D solids	Addition	Reading scales: weight	Understanding addition	Rounding to the nearest 100	Addition and subtraction: money	Multiplication	Identifying properties of 3-D solids	Number sequences	Subtraction	Handling data: Venn diagram	Identifying the reflection of a shape
AT	1 & 2	2	3	2	3	2	2	1 & 2	2	3	2	2	2	3
Question	1	2	3	4	5	6	7	8	9	10	11	12	13	14
Mark	1	1	1	1	1	1	1	1	1	1	1	1	1	1
Name														
1.														
2.														
3.														
4.														
5.														
6.														
7.														
8.														
9.														
10.														
11.														
12.														
13.														
14.														
15.														
16.														
17.														
18.														
19.														
20.														
21.														
22.														
23.														
24.														
25.														
26.														
27.														
28.														
29.														
30.														
Number correct														
Number incorrect or omitted														
Percentage correct														
Percentage incorrect or omitted														

Record-keeping format 3 Test 2 grid for test analysis (Paper B: Level 3)

	Halving	Relationship between units: weght	Handling data: bar chart	Reading scales: weght	Division	Position and movement	Understanding doubling and halving	Calculating time intervals	Subtraction	Explaining calculating methods	Finding three quarters of a shape	Number sequences: applying a rule	Identifying the unknown number in a number sentence	Adding numbers to make a multiple of 10	Properties of number: even numbers	Adding multiples of 10	Paper B score (out of 30)	**National Curriculum Level**
	2	3	2	3	1 & 2	3	1 & 2	3	2	2	2	1 & 2	1 & 2	2	1 & 2	1 & 2		
	15	16	17	18	19	20	21	22	23	24	25	26	27	28	29	30		
	1	1	1	1	1	1	1	1	1	1	1	1	1	1	1	1		
1.																		
2.																		
3.																		
4.																		
5.																		
6.																		
7.																		
8.																		
9.																		
10.																		
11.																		
12.																		
13.																		
14.																		
15.																		
16.																		
17.																		
18.																		
19.																		
20.																		
21.																		
22.																		
23.																		
24.																		
25.																		
26.																		
27.																		
28.																		
29.																		
30.																		

Record-keeping format 4 Test 3 grid for test analysis (Paper A: Level 2)

Name	Number sequence	Multiplication	Recognising 2-D shapes	Calculating (+, −, ×, ÷)	Recognising units of measure: capacity	Properties of number	Number sequences	Multiplication	Finding half of a shape	Identifying the correct number sentence	Rounding to the nearest 10	Handling data: table	Identifying the unknown number in a number sentence	Reading scales: capacity
AT	2	1 & 2	3	2	3	2	2	1 & 2	2	2	2	2	2	3
Question	1	2	3	4	5	6	7	8	9	10	11	12	13	14
Mark	1	1	1	1	1	1	1	1	1	1	1	1	1	1
1.														
2.														
3.														
4.														
5.														
6.														
7.														
8.														
9.														
10.														
11.														
12.														
13.														
14.														
15.														
16.														
17.														
18.														
19.														
20.														
21.														
22.														
23.														
24.														
25.														
26.														
27.														
28.														
29.														
30.														
Number correct														
Number incorrect or omitted														
Percentage correct														
Percentage incorrect or omitted														

Record-keeping format 4 Test 3 grid for test analysis (Paper A: Level 2)

	Understanding addition	Direction and movement	Properties of 3-D solids	Addition of more than two numbers	Addition: money	Doubling and addition / applying a rule	Reflective symmetry	Identifying the correct number sentence	Number sequences	Understanding multiplication and division	Rotation	Subtraction	Calculating time intervals	Division	Understanding multiplication and division	Understanding division / identifying unknown number in a number sentence
	2	3	3	2	1 & 2	1 & 2	3	1 & 2	2	2	3	2	3	2	2	2
	15	16	17	18	19	20	21	22	23	24	25	26	27	28	29	30
	1	1	1	1	1	1	1	1	1	1	1	1	1	1	1	1
1.																
2.																
3.																
4.																
5.																
6.																
7.																
8.																
9.																
10.																
11.																
12.																
13.																
14.																
15.																
16.																
17.																
18.																
19.																
20.																
21.																
22.																
23.																
24.																
25.																
26.																
27.																
28.																
29.																
30.																

Paper A score (out of 30)	**National Curriculum Level**

Record-keeping format 4 Test 3 grid for test analysis (Paper B: Level 3)

Name	Addition: money	Rounding to the nearest 100	Subtraction: capacity	Fractions / multiplication / division	Recognising 2-D shapes and right angles	Ordering whole numbers	Identifying unknown numbers in a number sentence	Handling data: bar chart	Understanding division / identifying unknown number in a number sentence	Properties of 3-D solids	Rounding to the nearest 10	Symmetrical patterns	Subtraction	Handling data: pictogram
AT	1 & 2	2	2 & 3	2	3	2	2	2	2	3	2	3	2	2
Question	1	2	3	4	5	6	7	8	9	10	11	12	13	14
Mark	1	1	1	1	1	1	1	1	1	1	1	1	1	1
1.														
2.														
3.														
4.														
5.														
6.														
7.														
8.														
9.														
10.														
11.														
12.														
13.														
14.														
15.														
16.														
17.														
18.														
19.														
20.														
21.														
22.														
23.														
24.														
25.														
26.														
27.														
28.														
29.														
30.														
Number correct														
Number incorrect or omitted														
Percentage correct														
Percentage incorrect or omitted														

Record-keeping format 4 Test 3 grid for test analysis (Paper B: Level 3)

	Identifying a number on a number line	Rotation through a right angle	Finding three quarters of a shape	Number sequences	Properties of 2-D shapes / identifying right angles	Division: length	Explaining calculating methods	Multiplication	Capacity: reading scales / subtraction	Calculating time intervals	Identifying unknown number in a number sentence	Division: rounding after division	Addition	Multiplication and addition: money	Halving (inverse)	Identifying the rule of a number sequence	Paper B score (out of 30)	**National Curriculum Level**
	2	3	2	2	3	1, 2 & 3	1 & 2	2	3	3	2	1 & 2	2	1 & 2	2	1 & 2		
	15	16	17	18	19	20	21	22	23	24	25	26	27	28	29	30		
	1	1	1	1	1	1	1	1	1	1	1	1	1	1	1	1		
1.																		
2.																		
3.																		
4.																		
5.																		
6.																		
7.																		
8.																		
9.																		
10.																		
11.																		
12.																		
13.																		
14.																		
15.																		
16.																		
17.																		
18.																		
19.																		
20.																		
21.																		
22.																		
23.																		
24.																		
25.																		
26.																		
27.																		
28.																		
29.																		
30.																		

Record-keeping format 5 Attainment Target 1 – Using and applying mathematics

Performance criteria (P4–P8)

P4	P5	P6	P7	P8
• Is aware of cause and effect in familiar mathematical activities. • Shows awareness of changes in shape, position or quantity. • Anticipates, follows and joins in familiar mathematical activities when given a contextual cue.	• With support, matches objects or pictures. • Beginning to sort sets of objects, according to a single attribute. • Makes sets that have the same small number of objects in each. • Solves simple problems practically.	• Sorts objects and materials according to given criteria. • Beginning to identify when an object is different and does not belong to given categories. • Copies simple patterns or sequences.	• Completes a range of classification activities using given criteria. • Identifies when an object is different and does not belong to a given familiar category.	• Recognises, describes and recreates simple repeating patterns and sequences. • Beginning to use their developing mathematical understanding of counting to solve simple problems they may encounter in play, games or other work. • Beginning to make simple estimates, such as how many cubes will fit in a box.

Level 1

Problem solving	Communicating	Reasoning
• Use mathematics as an integral part of classroom activities • Use materials for a practical task	• Represent work with objects or pictures and discuss it	• Recognise and use simple patterns or relationships

Level 2

Problem solving	Communicating	Reasoning
• Select and use material in some classroom activities • Select and use mathematics for some classroom activities • Begin to develop own strategies for solving a problem • Begin to understand ways of working through a problem	• Discuss work using mathematical language • Respond to and ask mathematical questions • Begin to represent work using symbols and simple diagrams • Explain why an answer is correct	• Ask questions such as: 'What would happen if...?' 'Why?' • Begin to develop simple strategies

Level 3

Problem solving	Communicating	Reasoning
• Develop different mathematical approaches to a problem • Look for ways to overcome difficulties • Begin to make decisions and realise that results may vary according to the 'rule' used • Begin to organise work • Check results	• Discuss mathematical work • Begin to explain thinking • Use and interpret mathematical symbols and diagrams	• Understand a general statement • Investigate general statements and predictions by finding and trying out examples

General comments

Year 2 National Expectations
Start-of-year: Level 1a (2c)
End-of-year: Level 2b

Record-keeping format 6 Attainment Target 2 – Number and algebra

Performance criteria (P4–P8)

P4	P5	P6	P7	P8
• Shows an interest in number activities and counting.	• Responds to and joins in with familiar number rhymes, stories, songs and games. • Can indicate 'one' or 'two'. • Demonstrates that they are aware of contrasting quantities, by making groups of objects with help.	• Demonstrates an understanding of one-to-one correspondence in a range of contexts. • Joins in rote counting up to five and uses numbers to five in familiar contexts and games. • Counts reliably to three and makes sets of up to three objects. • Demonstrates an understanding of the concept of more/fewer. • Uses 1p coins. • Joins in with new number rhymes, songs, stories and games with some assistance or encouragement.	• Joins in rote counting to ten. • Counts at least five objects reliably. • Beginning to recognise numerals from 1 to 5 and to understand that each represents a constant number or amount. • Responds appropriately to key vocabulary and questions. • Beginning to recognise differences in quantity. • In practical situations responds to 'add one' and 'take one'.	• Joins in with rote counting beyond ten. • Continues the rote count onwards from a given small number. • Beginning to count up to ten objects. • Compares two given numbers of objects saying which is more and which is less. • Beginning to recognise numerals from 1 to 9 and relate them to sets of objects. • In practical situations can add one to or take one away from a number of objects. • Beginning to use ordinal numbers when describing the position of objects, people or events. • Estimates a small number and checks by counting.

Level 1*

1C	1B	1A
• Reads most numbers up to 10 in familiar contexts. • Makes an attempt to record numbers up to 10. • In practical situations begins to use the vocabulary involved in addition and subtraction. • Demonstrates an understanding of addition as the combining of two or more groups of objects. • Demonstrates an understanding of subtraction as the taking away of objects from a group.	• Can count, read and order numbers (including using ordinal numbers) up to 10 in a range of settings. • Writes numerals up to 10 with increasing accuracy. • Using numbers up to 10, solves problems involving addition and subtraction, including comparing two sets to find a numerical difference.	• Can count, read and order numbers from 0 to 20. • Records numbers from 0 to 10 and associates these with the number of objects they have counted. • Recognises 0 as 'none' and 'zero' in stories and rhymes and when counting and ordering. • Understands the operation of addition and subtraction and uses the related vocabulary. • Can add and subtract numbers when solving problems involving up to 10 objects in a range of contexts.

Level 2*

2C	2B	2A
• Is confident in using numbers up to 20. • Is beginning to understand place value. • Is beginning to know by heart all pairs of whole numbers with totals up to 10 and can use these facts to add or subtract a pair of numbers mentally. • Recognises odd and even numbers to 20. • Recognises other simple number sequences. • Collects data by counting and records the data in a tally or block chart.	• Can count, read, write and order accurately whole numbers to at least 50. • Recognises that subtraction is the inverse of addition and uses this to solve addition and subtraction problems. • Identifies doubles and halves using numbers up to 20. • Is beginning to understand the concept of 'a quarter'. • Recognises odd and even numbers to about 50. • Recognises 1p, 2p, 5p, 10p, 20p and 50p coins. • Chooses coins to make amounts up to 50p. • Organises and classifies data using simple lists and tables.	• Can count, read, write and order whole numbers to at least 100. • Uses mental recall of addition facts up to 10 to add and subtract whole numbers, including multiples of 10. • Understands the operation of multiplication as repeated addition and as a way of representing the number of items in a rectangular array. • Understands the operation of division as repeated subtraction or sharing. • Understands halving as the inverse of doubling and uses this to derive and learn multiplication and division facts from the 2 times table. • Understands and uses £ p notation for money. • Sorts objects and classifies them using more than one criterion. • Presents data that has been collected in simple lists, tables or block graphs and communicates their findings to others.

Level 3

Numbers and the number system	Calculations	Solving numerical problems	Processing, representing and interpreting data
• Understand place value (ThHTU) • Begin to use decimal notation • Recognise negative numbers • Use simple fractions that are several parts of a whole • Recognise when two fractions are equivalent	• Make approximations • Recall addition and subtraction number facts to 20 • Add and subtract two two-digit numbers mentally • Add and subtract three two-digit numbers using written methods • Recall 2, 3, 4, 5, 10 multiplication tables • Recall division facts corresponding to the 2, 3, 4, 5, 10 multiplication tables	• Solve word problems involving larger numbers • Solve word problems involving multiplication • Solve word problems involving division, including those with a remainder	• Extract and interpret information presented in simple tables and lists • Construct and interpret bar charts and pictograms where the symbol represents a group of units

General comments

Year 2 National Expectations
Start-of-year: Level 1a (2c)
End-of-year: Level 2b

* The breakdown of Levels 1 and 2 into A, B, C has been taken from the DfES publication *Supporting the Target Setting Process*, 2001.

Record-keeping format 7 Attainment Target 3 – Shape, space and measures

Performance criteria (P4–P8)

P4	P5	P6	P7	P8
• Beginning to search for objects that have gone out of sight, hearing or touch, demonstrating the beginning of object permanence. • Demonstrates an interest in position and the relationship between objects.	• Searches intentionally for objects in their usual places. • Compares the overall size of one object with that of another where there is a marked difference. • Finds bigger and smaller objects on request. • Explores the position of objects.	• Searches for objects not found in their usual place demonstrating their understanding of object permanence. • Compares the overall size of one object with that of another where the difference is not great. • Manipulates three-dimensional shapes. • Shows understanding of words, signs and symbols that describe position. • Shows awareness of vocabulary such as 'more' and 'less', in practical situations.	• Responds to forwards and backwards. • Starts to pick out named shapes from a collection. • Uses familiar words when they compare sizes and quantities and describe position.	• Compares directly, two length or heights where the difference is marked and can indicate 'the long one' or 'the tall one'. • Shows awareness of time, through some familiarity with the names of the days of the week and significant times in their day, such as meal times, bed times. • Beginning to use mathematical vocabulary such as 'straight', 'circle', 'larger' to describe the shape and size of solids and flat shapes. • Describes shapes in simple models, pictures and patterns.

Level 1*

1C	1B	1A
• Constructs with 3-D shapes. • Makes arrangements and patterns of 2-D shapes. • Recognises and names some familiar 2-D shapes such as circle, triangle and square. • Matches and sorts shapes in activities. • Is beginning to use their knowledge of shape to describe the properties of everyday objects, and to compare them by size. • Uses everyday language to describe position. • Uses everyday language to compare two quantities.	• Works with, recognises and names common 3-D shapes such as cube and cylinder and 2-D shapes such as circle, triangle, rectangle and square. • Describes the basic properties of shapes. • Makes simple comparisons between shapes using such terms as 'larger', 'smaller', curved', 'straight'. • Recognises terms describing position such as 'behind', 'in front of' and 'on top'. • Measures and orders more than two objects (by length, weight or capacity), using direct comparison. • Orders everyday events logically. • Beginning to use the vocabulary of time.	• Sorts and describes 3-D and 2-D shapes in terms of their properties and positions. • Compares two lengths, weights or capacities by direct comparison. • Continues and creates simple spatial patterns. • Recognises directional symbols such as arrows.

Level 2*

2C	2B	2A
• Uses the correct terms for common shapes, e.g. circle, triangle, cube, cylinder. • Describes the properties of 3-D and 2-D shapes using everyday language. • Is beginning to link everyday language with mathematical language. • Suggests suitable units and measuring equipment to estimate and measure a length, mass or capacity.	• Uses correct terms for common shapes. • Recognises properties such as faces, edges, sides and corners. • Distinguishes between straight and turning movements. • Describes positions using terms such as 'at the corner of', 'further away from'. • Recognises and draws a line of symmetry or constructs patterns with a line of symmetry. • Beginning to make simple measurements of length, weight and capacity accurately. • Becoming familiar with using standard units of measurements.	• Identifies common shapes by their properties. • Describes shapes in terms of their properties, including recognising right angles in 2-D and 3-D shapes. • Sorts on collection of 2-D or 3-D shapes in more than one way. • Identifies lines of symmetry in simple shapes and recognises shapes with no lines of symmetry. • Beginning to understand angle as a measure of turn. • Shows an understanding of right angles through movement, including using clockwise and anticlockwise. • Tells the time using hour, half-hour and quarter-hour units. • Uses the vocabulary related to time. • Beginning to use standard units of length (cm, m), mass (g, kg), and capacity (litres) to measure and compare quantities and objects. • Compares events and timescales using appropriate standard units of time (hour, minute, second).

Level 3

Understanding properties of shapes	**Understanding properties of position and movement**	**Understanding measures**
• Classify 3-D and 2-D shapes in various ways	• Use mathematical properties such as reflective symmetry to describe 2-D shapes	• Use non-standard units of length, capacity and mass in a range of contexts • Use standard units of length, capacity, mass and time in a range of contexts

General comments

Year 2 National Expectations
Start-of-year: Level 1a (2c)
End-of-year: Level 2b

* The breakdown of Levels 1 and 2 into A, B, C has been taken from the DfES publication *Supporting the Target Setting Process*, 2001.

Record-keeping format 8 Class record of the end-of-year expectations

Class: ______________________

Date: ______________________

Year 2 **End-of-year expectations**	**Names**																													
Counting and understanding number (AT2) Count up to 100 objects by grouping them and counting in tens, fives or twos; explain what each digit in a two-digit number represents, including numbers where 0 is a place holder; partition two-digit numbers in different ways, including into multiples of 10 and 1 (Level 2)																														
Knowing and using number facts (AT2) Derive and recall all addition and subtraction facts for each number to at least 10, all pairs with totals to 20 and all pairs of multiples of 10 with totals up to 100 (Level 2)																														
Calculating (AT2) Add or subtract mentally a one-digit number or a multiple of 10 to or from any two-digit number; use practical and informal written methods to add and subtract two-digit numbers (Level 2)																														
Use the symbols +, −, ×, ÷ and = to record and interpret number sentences involving all four operations; calculate the value of an unknown in a number sentence (Level 2)																														
Understanding shape (AT3) Visualise common 2-D shapes and 3-D solids; identify shapes from pictures of them in different positions and orientations; sort, make and describe shapes, referring to their properties (Level 2)																														
Measuring (AT3) Use units of time (seconds, minutes, hours, days) and know the relationships between them; read the time to the quarter hour; identify time intervals, including those that cross the hour (Level 2)																														
Handling data (AT2) Use lists, tables and diagrams to sort objects; explain choices using appropriate language, including 'not' (Level 2)																														

NOTES: **Using and applying mathematics (AT1)** is incorporated throughout
End-of-year National Expectations: Level 2b

Record-keeping format 9 Individual child's record of the end-of-year expectations

Name: ______________________

Foundation Stage		Year 1		Year 2		Year 3	
Using and applying mathematics (AT1)							
Use developing mathematical ideas and methods to solve practical problems (Level 1)							
Talk about, recognise and recreate simple patterns (Level 1)							
Counting and understanding number (AT2)							
Say and use the number names in order in familiar contexts (Level 1)		Read and write numerals from 0 to 20, then beyond; use knowledge of place value to position these numbers on a number track and number line (Level 2)		Count up to 100 objects by grouping them and counting in tens, fives or twos; explain what each digit in a two-digit number represents, including numbers where 0 is a place holder; partition two-digit numbers in different ways, including into multiples of 10 and 1 (Level 2)		Partition three-digit numbers into multiples of 100, 10 and 1 in different ways (Level 3)	
Count reliably up to 10 everyday objects (Level 1)							
Use language such as 'more' or 'less' to compare two numbers (Level 1)							
Recognise numerals 1 to 9 (Level 1)							
Knowing and using number facts (AT2)							
Find one more or one less than a number from 1 to 10 (Level 1)		Derive and recall all pairs of numbers with a total of 10 and addition facts for totals to at least 5; work out the corresponding subtraction facts (Level 2)		Derive and recall all addition and subtraction facts for each number to at least 10, all pairs with totals to 20 and all pairs of multiples of 10 with totals up to 100 (Level 2)		Derive and recall all addition and subtraction facts for each number to 20, sums and differences of multiples of 10 and number pairs that total 100 (Level 3)	
Calculating (AT2)							
Begin to relate addition to combining two groups of objects and subtraction to 'taking away' (Level 1)		Use the vocabulary related to addition and subtraction and symbols to describe and record addition and subtraction number sentences (Level 2)		Add or subtract mentally a one-digit number or a multiple of 10 to or from any two-digit number; use practical and informal written methods to add and subtract two-digit numbers (Level 2)		Add or subtract mentally combinations of one-digit and two-digit numbers (Level 3)	
In practical activities and discussion begin to use the vocabulary involved in adding and subtracting (Level 1)				Use the symbols +, −, ×, ÷ and = to record and interpret number sentences involving all four operations; calculate the value of an unknown in a number sentence (Level 2)			
Understanding shape (AT3)							
Use language such as 'circle' or 'bigger' to describe the shape and size of solids and flat shapes (Level 1)		Visualise and name common 2-D shapes and 3-D solids and describe their features; use them to make patterns, pictures and models (Level 2)		Visualise common 2-D shapes and 3-D solids; identify shapes from pictures of them in different positions and orientations; sort, make and describe shapes, referring to their properties (Level 2)		Draw and complete shapes with reflective symmetry and draw the reflection of a shape in a mirror line along one side (Level 3)	
Use everyday words to describe position (Level 1)							
Measuring (AT3)							
Use language such as 'greater', 'smaller', 'heavier' or 'lighter' to compare quantities (Level 1)		Estimate, measure, weigh and compare objects, choosing and using suitable uniform non-standard or standard units and measuring instruments, e.g. a lever balance, metre stick or measuring jug (Level 2)		Use units of time (seconds, minutes, hours, days) and know the relationships between them; read the time to the quarter hour; identify time intervals, including those that cross the hour (Level 2)		Read, to the nearest division and half-division, scales that are numbered or partially numbered; use the information to measure and draw to a suitable degree of accuracy (Level 3)	
		Handling data (AT4)					
		Answer a question by recording information in lists and tables; present outcomes using practical resources, pictures, block graphs or pictograms (Level 2)		Use lists, tables and diagrams to sort objects; explain choices using appropriate language, including 'not' (Level 2)		Use Venn diagrams or Carroll diagrams to sort data and objects using more than one criterion (Level 3)	

NOTE: **Using and applying mathematics (AT1)** is incorporated throughout

Record-keeping format 9 Individual child's record of the end-of-year expectations

Name: ____________________

Year 4		Year 5		Year 6		Year 6 progression to Year 7	
Counting and understanding number (AT2)							
Use diagrams to identify equivalent fractions; interpret mixed numbers and position them on a number line (Level 3)		Explain what each digit represents in whole numbers and decimals with up to two places, and partition, round and order these numbers (Level 4)		Express one quantity as a percentage of another; find equivalent percentages, decimals and fractions (Level 4)		Use ratio notation, reduce a ratio to its simplest form and divide a quantity into two parts in a given ratio; solve simple problems involving ratio and direct proportion (Level 5)	
Knowing and using number facts (AT2)							
Derive and recall multiplication facts up to 10 × 10, the corresponding division facts and multiples of numbers to 10 up to the tenth multiple (Level 4)		Use knowledge of place value and addition and subtraction of two-digit numbers to derive sums and differences and doubles and halves of decimals (Level 4)		Use knowledge of place value and multiplication facts to 10 × 10 to derive related multiplication and division facts involving decimals (Level 4)		Make and justify estimates and approximations to calculations (Level 5)	
Calculating (AT2)							
Add or subtract mentally pairs of two-digit whole numbers (Level 3)		Use efficient written methods to add and subtract whole numbers and decimals with up to two places (Level 4)		Use efficient written methods to add and subtract integers and decimals, to multiply and divide integers and decimals by a one-digit integer, and to multiply two-digit and three-digit integers by a two-digit integer (Level 5)		Use bracket keys and the memory of a calculator to carry out calculations with more than one step; use the square-root key (Level 5)	
Develop and use written methods to record, support and explain multiplication and division of two-digit numbers by a one-digit number, including division with remainders (Level 4)							
Understanding shape (AT3)							
Know that angles are measured in degrees and that one whole turn is 360°; compare and order angles less than 180° (Level 3)		Read and plot co-ordinates in the first quadrant; recognise parallel and perpendicular lines in grids and shapes; use a set-square and ruler to draw shapes with perpendicular or parallel sides (Level 4)		Visualise and draw on grids of different types where a shape will be after reflection, after translations, or after rotation through 90° or 180° about its centre or one of its vertices (Level 5)		Know the sum of angles on a straight line, in a triangle and at a point, and recognise vertically opposite angles (Level 5)	
Measuring (AT3)							
Choose and use standard metric units and their abbreviations when estimating, measuring and recording length, weight and capacity; know the meaning of 'kilo', 'centi' and 'milli' and, where appropriate, use decimal notation to record measurements (Level 3)		Draw and measure lines to the nearest millimetre; measure and calculate the perimeter of regular and irregular polygons; use the formula for the area of a rectangle to calculate the rectangle's area (Level 4)		Select and use standard metric units of measure and convert between units using decimals to two places (Level 4)		Solve problems by measuring, estimating and calculating; measure and calculate using imperial units still in everyday use; know their approximate metric values (Level 5)	
Handling data (AT4)							
Answer a question by identifying what data to collect; organise, present, analyse and interpret the data in tables, diagrams, tally charts, pictograms and bar charts, using ICT where appropriate (Level 3)		Construct frequency tables, pictograms and bar and line graphs to represent the frequencies of events and changes over time (Level 4)		Solve problems by collecting, selecting, processing, presenting and interpreting data, using ICT where appropriate; draw conclusions and identify further questions to ask (Level 5)		Understand and use the probability scale from 0 to 1; find and justify probabilities based on equally likely outcomes in simple contexts (Level 5)	

NOTE: **Using and applying mathematics (AT1)** is incorporated throughout

	Foundation Stage	Year 1	Year 2	Year 3	Year 4	Year 5	Year 6
End-of-year National Expectations	1b	1a (2c)	2b	2a (3c)	3b	3a (4c)	4b

Word problem cards

1. 16 children are playing football. 4 more join them. How many children are now playing football?

2. Lisa has £1. She spends half her money on a chocolate bar. How much money does she have left?

3. There are 30 children in 2A and 31 children in 2B. How many children are there altogether in Year 2?

4. Carol has a piece of ribbon 30 cm long. She cuts the ribbon in half. How long is each piece of ribbon now?

5. For a party, Fred mixes together 6 litres of juice and 2 litres of water. How many litres is this altogether?

6. There are 18 children in the pool. 5 more jump in. How many children are now in the pool?

7. There are 16 pencils. 8 children share them equally. How many pencils does each child have?

8. Tom walks to school. He leaves home at 8:30 and arrives at school at 8:45. How long does it take Tom to walk to school?

9. Mike weighs 16 kg. Jason is 3 kg lighter than Mike. How heavy is Jason?

10. A pet shop has 28 fish in a tank. 5 fish are sold. How many fish are left in the tank?

11. There are 40 cows in a field. 20 cows are black and the rest are brown. How many brown cows are there?

12. Jane blows up 20 balloons. 6 of them burst. How many balloons are left?

13. 8 children have 10 fingers each. How many fingers is this altogether?

14. Eggs come in cartons of 6. How many eggs are there in 5 cartons?

15. Mr. West has 6 shirts. Each shirt has 4 small buttons and 6 large buttons. How many buttons are there altogether on his 6 shirts?

16. I think of a number, then halve it. The answer is 7. What was my number?

17. Jane has £5. Sam has twice as much money as Jane, but spends £2. How much money does Sam now have?

18. There are 25 marbles in a bag. Jake takes 12 and Freda takes 6. How many marbles are left?

19. 25 ducks are in the pond. 12 fly away. How many ducks are left in the pond?

20. On Monday, 7 children had an ice cream on their way home from school. On Friday, twice as many had an ice cream. How many children had an ice cream on Monday and Friday?

Puzzles 1

1.

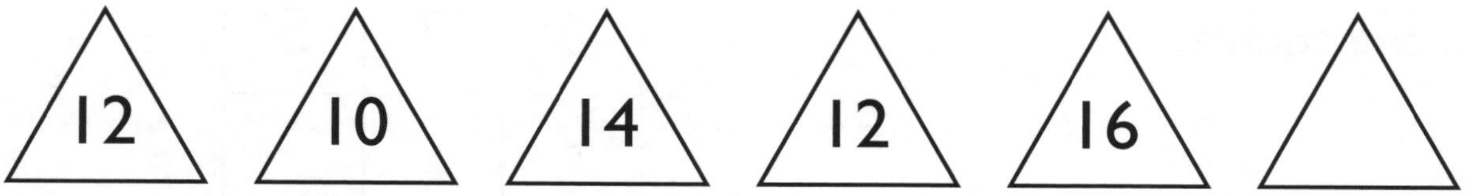
12
10
14
12
16

What is the next number in the sequence?

2.

What time does the last clock show?

3.

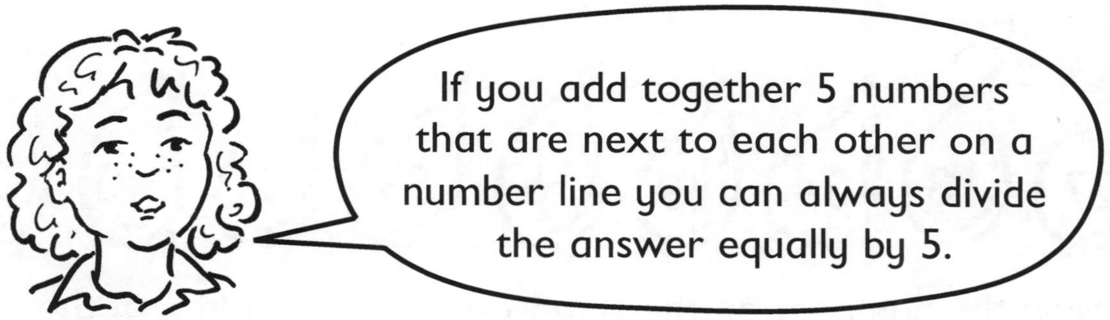
If you add together 5 numbers that are next to each other on a number line you can always divide the answer equally by 5.

- Is Tracey right?

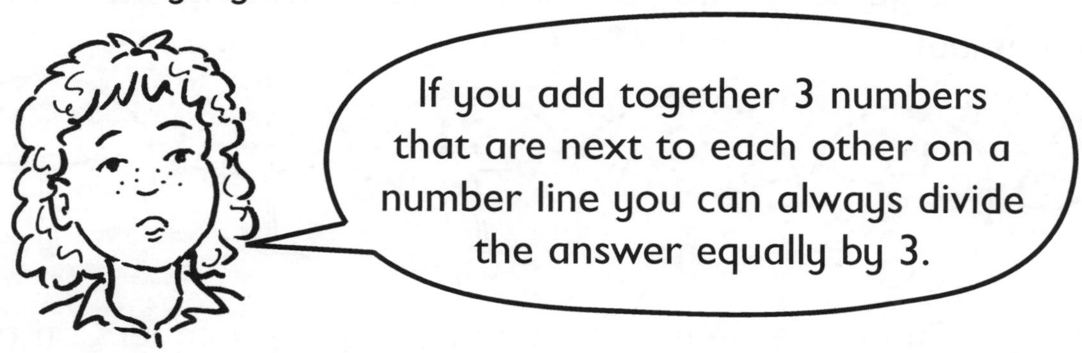
If you add together 3 numbers that are next to each other on a number line you can always divide the answer equally by 3.

- Is Tracey right this time?

Puzzles 2

4. Write the missing numbers in the square.

6	7	5
5		
7		6

5. Complete the shapes to make them symmetrical.

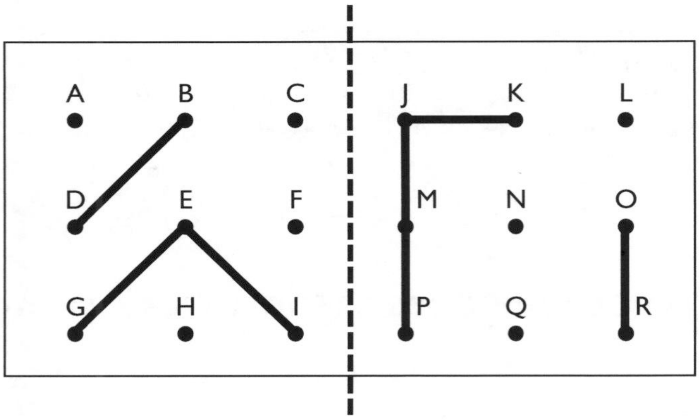

6.

- Using the numbers on the raindrops, can you add 2 of the numbers together or find the difference between them to make the numbers on the umbrellas?

- Can you find more than one way of using the numbers on the raindrops to make the numbers on the umbrellas?

1–20 number cards

1	2	3	4
5	6	7	8
9	10	11	12
13	14	15	16
17	18	19	20

Two-digit number cards

22	26	33	35
38	41	44	52
57	60	67	69
73	79	81	85
86	92	94	98

Three-digit number cards

113	169	210	258
320	397	474	481
522	582	603	654
701	735	736	845
846	897	967	978

Four-digit number cards

1240	1481	1532	2693
2865	3024	3156	4379
4718	5084	5499	5907
6141	6623	7515	7960
8282	8734	9356	9877

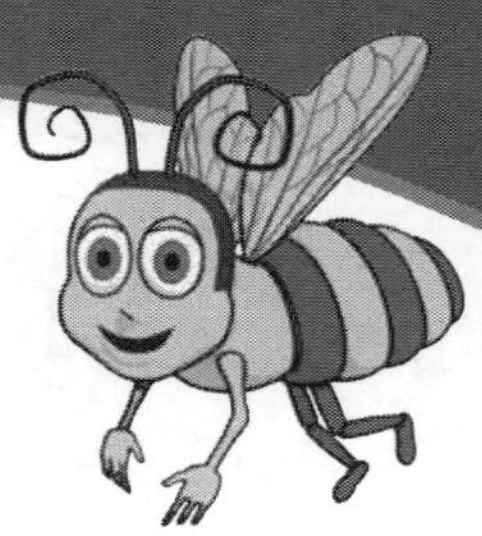

Continuing number sequences

6, 8, 10, 12, 14, …

37, 34, 31, 28, 25, …

4, 9, 14, 19, 24, …

152, 142, 132, 122, 112, …

3, 7, 11, 15, 19, …

8, 11, 14, 17, 20, …

88, 83, 78, 73, 68, …

54, 64, 74, 84, 94, …

46, 44, 42, 40, 38, …

51, 47, 43, 39, 35, …

Symbol cards

<	<	<	<
<	<	<	<
=	=	=	=
=	=	=	=
□	□	□	□

Round to 10

Collins New Primary Maths

Fractions of shapes cards

A

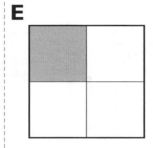

B

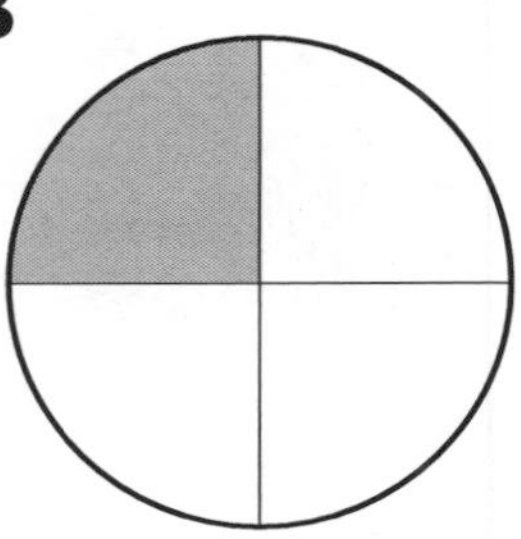

C

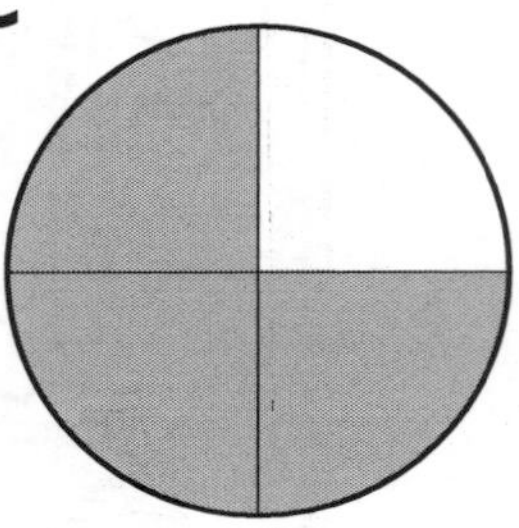

D

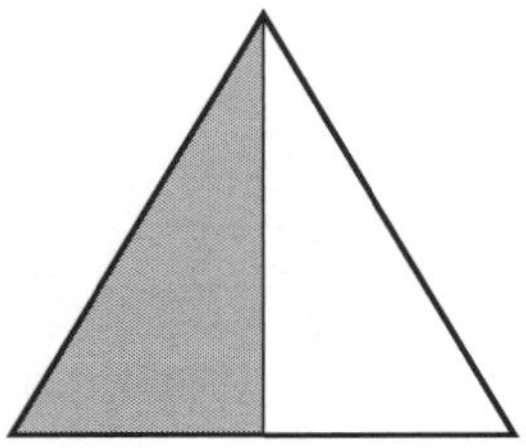

E

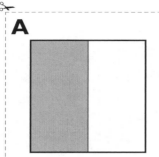

F

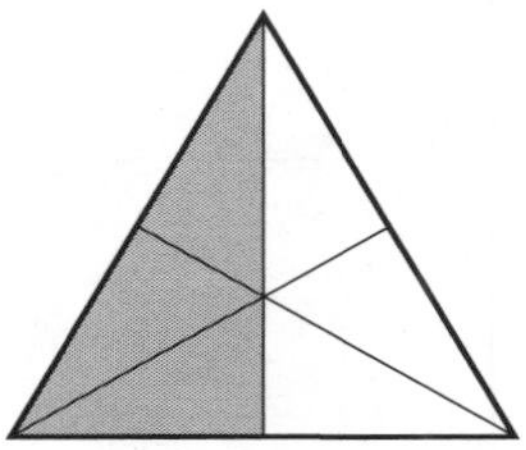

G

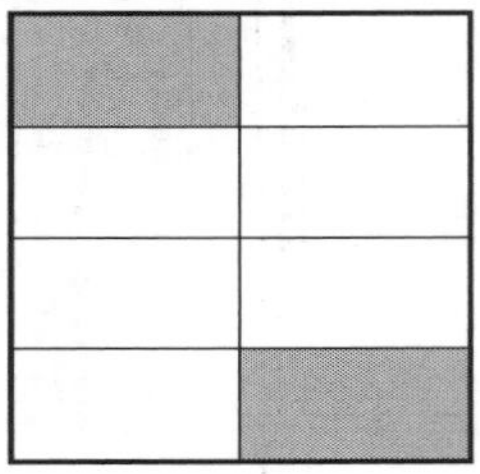

H

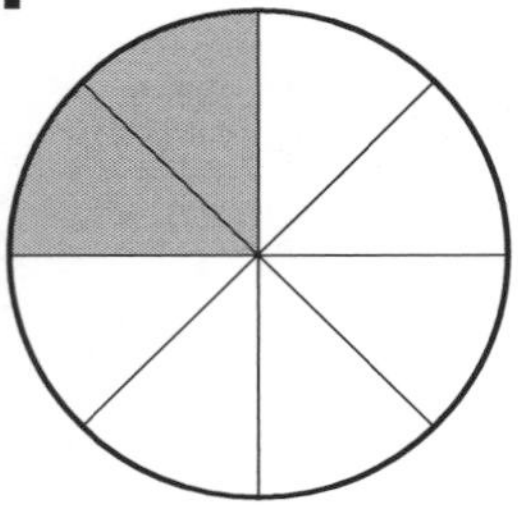

I

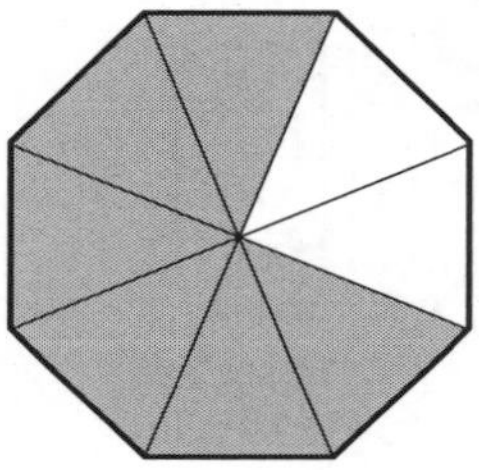

J

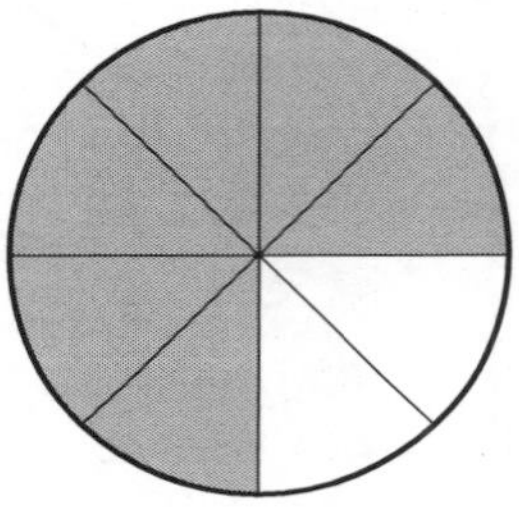

K

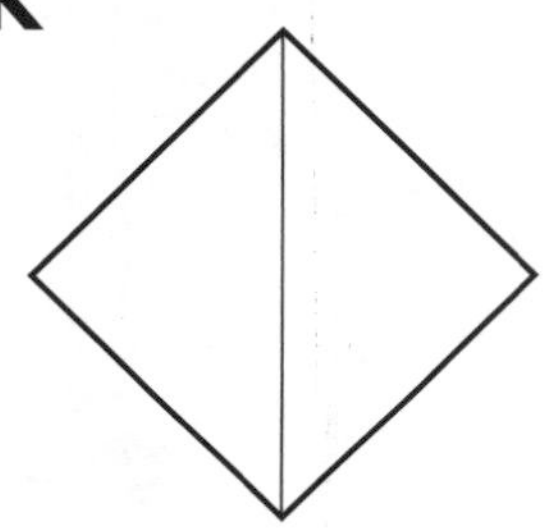

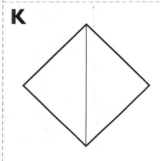

L

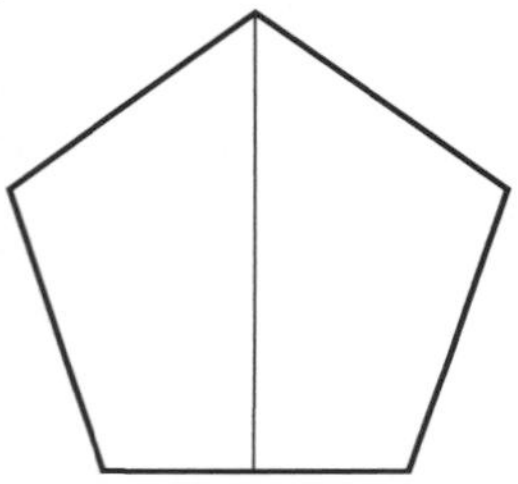

M

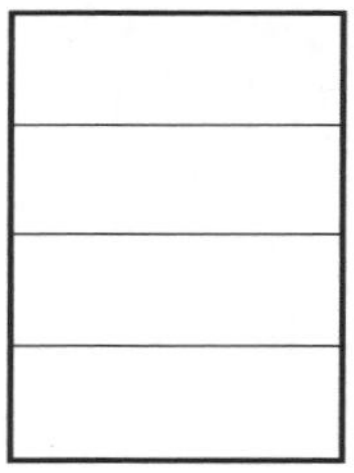

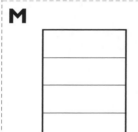

N

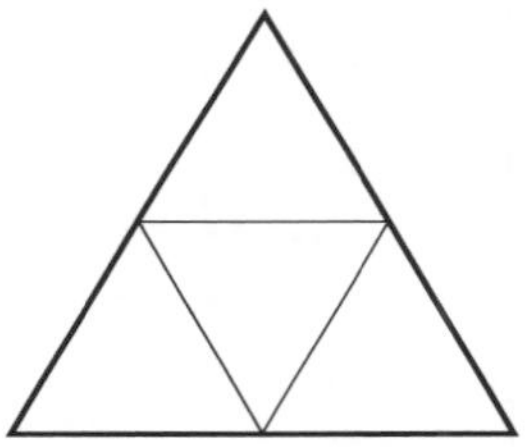

O

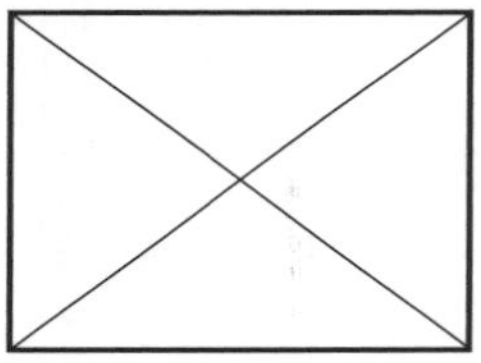

P

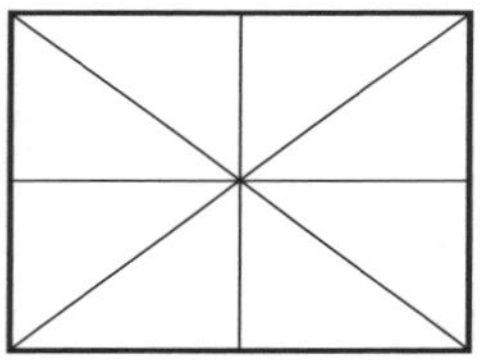

Q

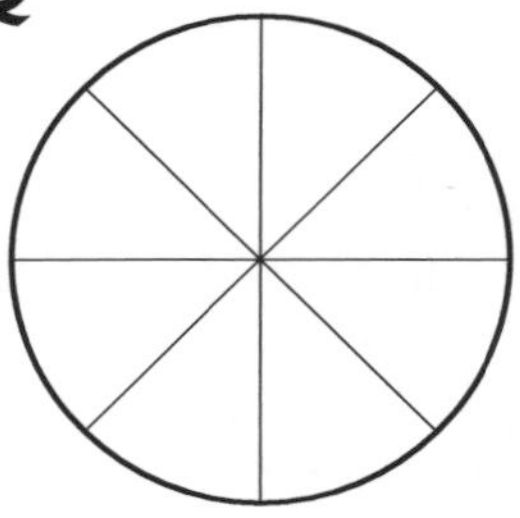

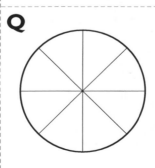

R

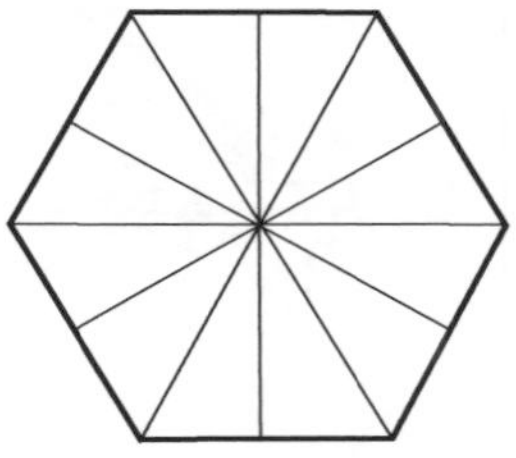

S

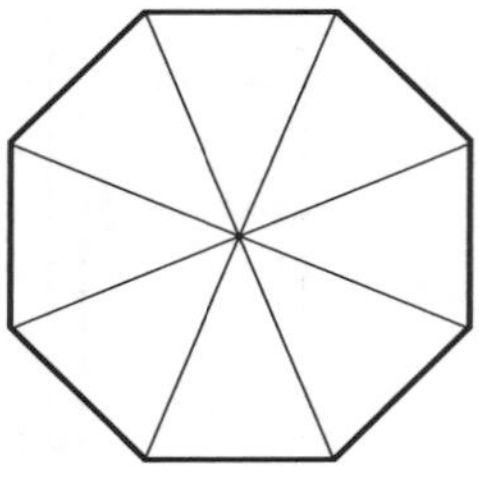

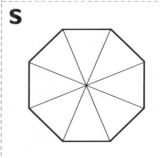

T

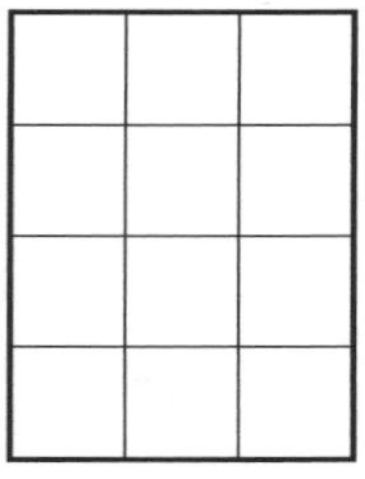

Collins New Primary Maths

Array cards

A

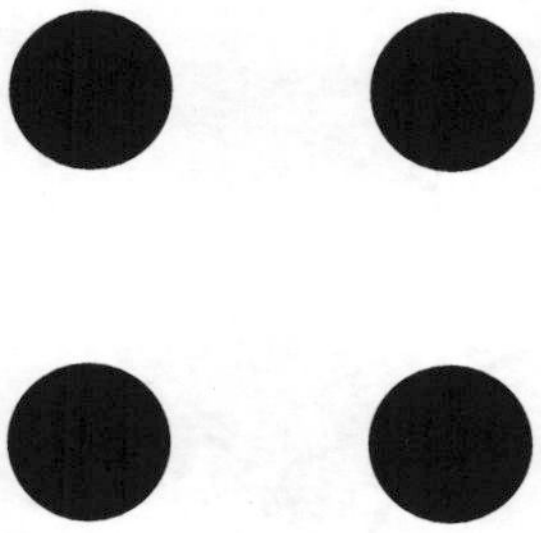

B

C

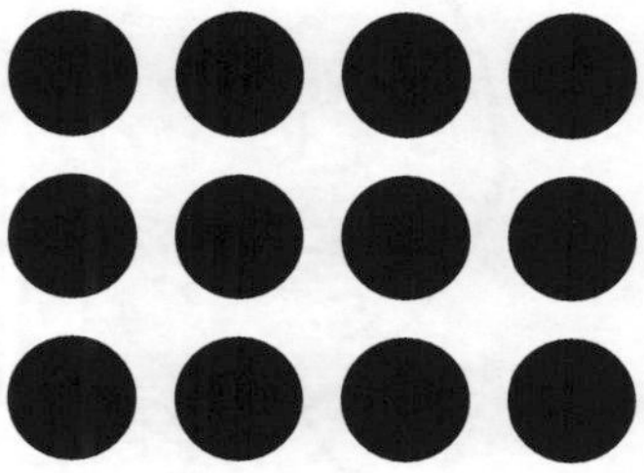

D

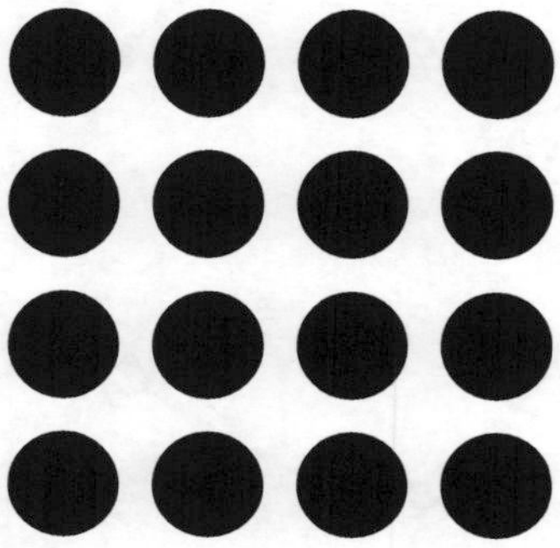

E

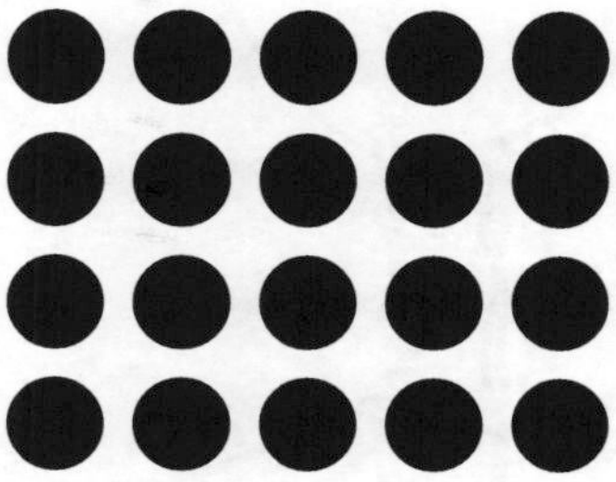

F

G

H

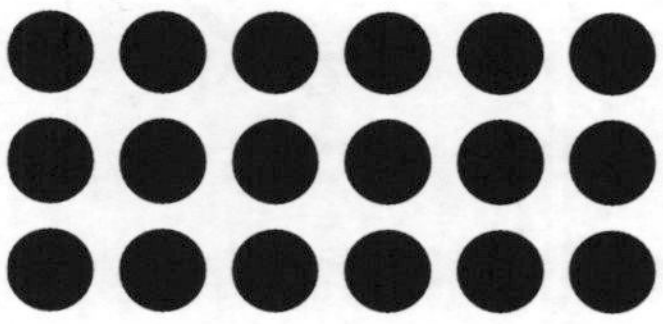

I

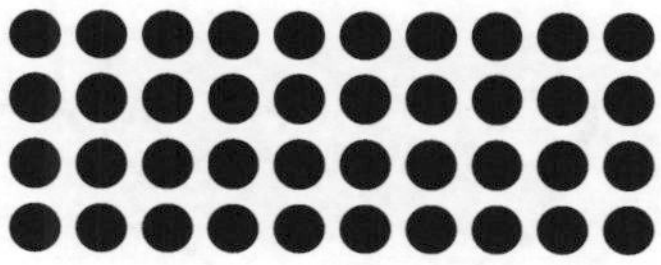

J

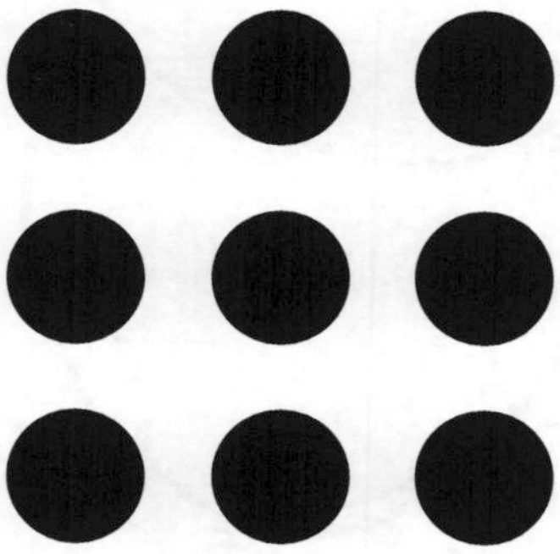

K

L

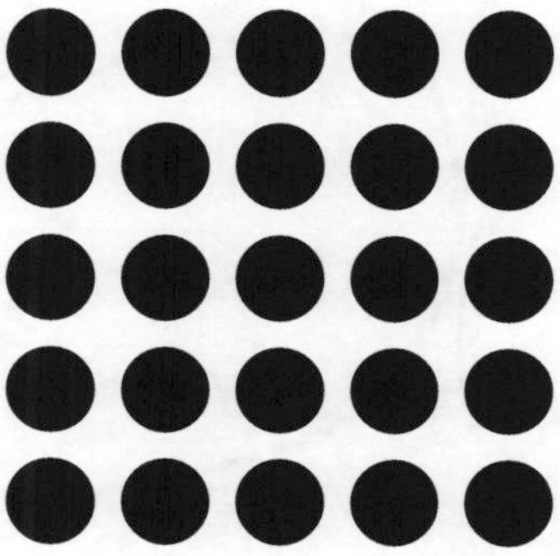

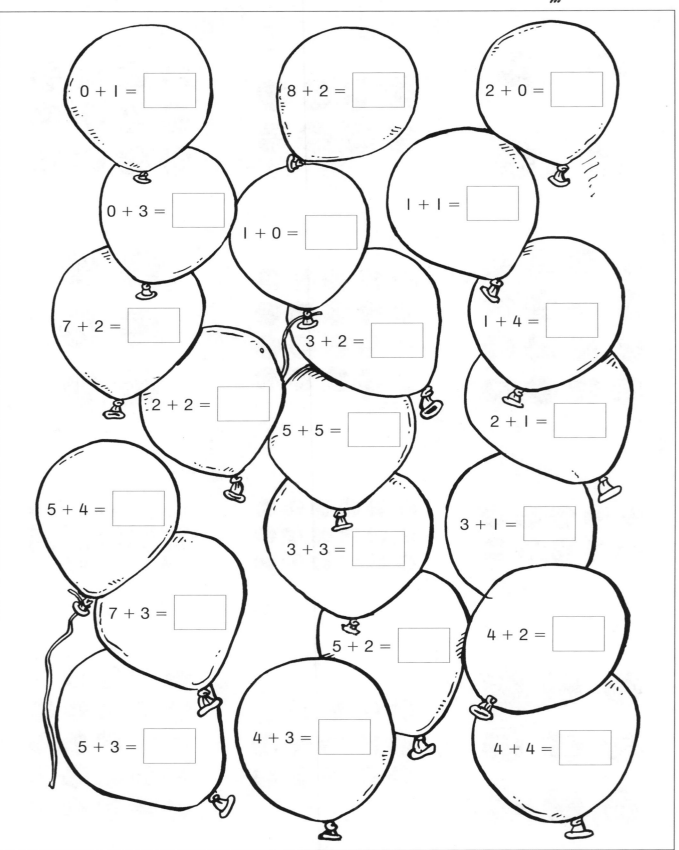

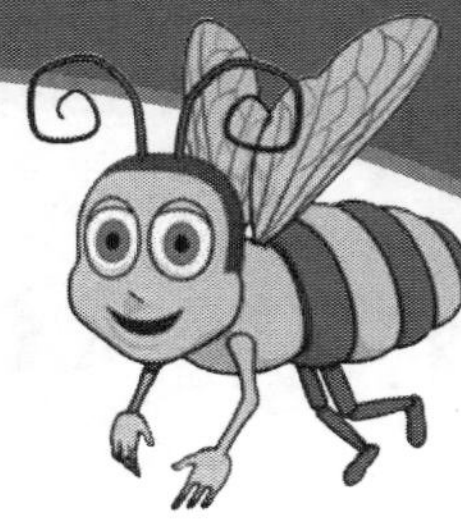

Addition facts to 10

0 + 1 = ☐

8 + 2 = ☐

2 + 0 = ☐

0 + 3 = ☐

1 + 0 = ☐

1 + 1 = ☐

7 + 2 = ☐

3 + 2 = ☐

1 + 4 = ☐

2 + 2 = ☐

5 + 5 = ☐

2 + 1 = ☐

5 + 4 = ☐

3 + 3 = ☐

3 + 1 = ☐

7 + 3 = ☐

5 + 2 = ☐

4 + 2 = ☐

5 + 3 = ☐

4 + 3 = ☐

4 + 4 = ☐

Subtraction facts to 10

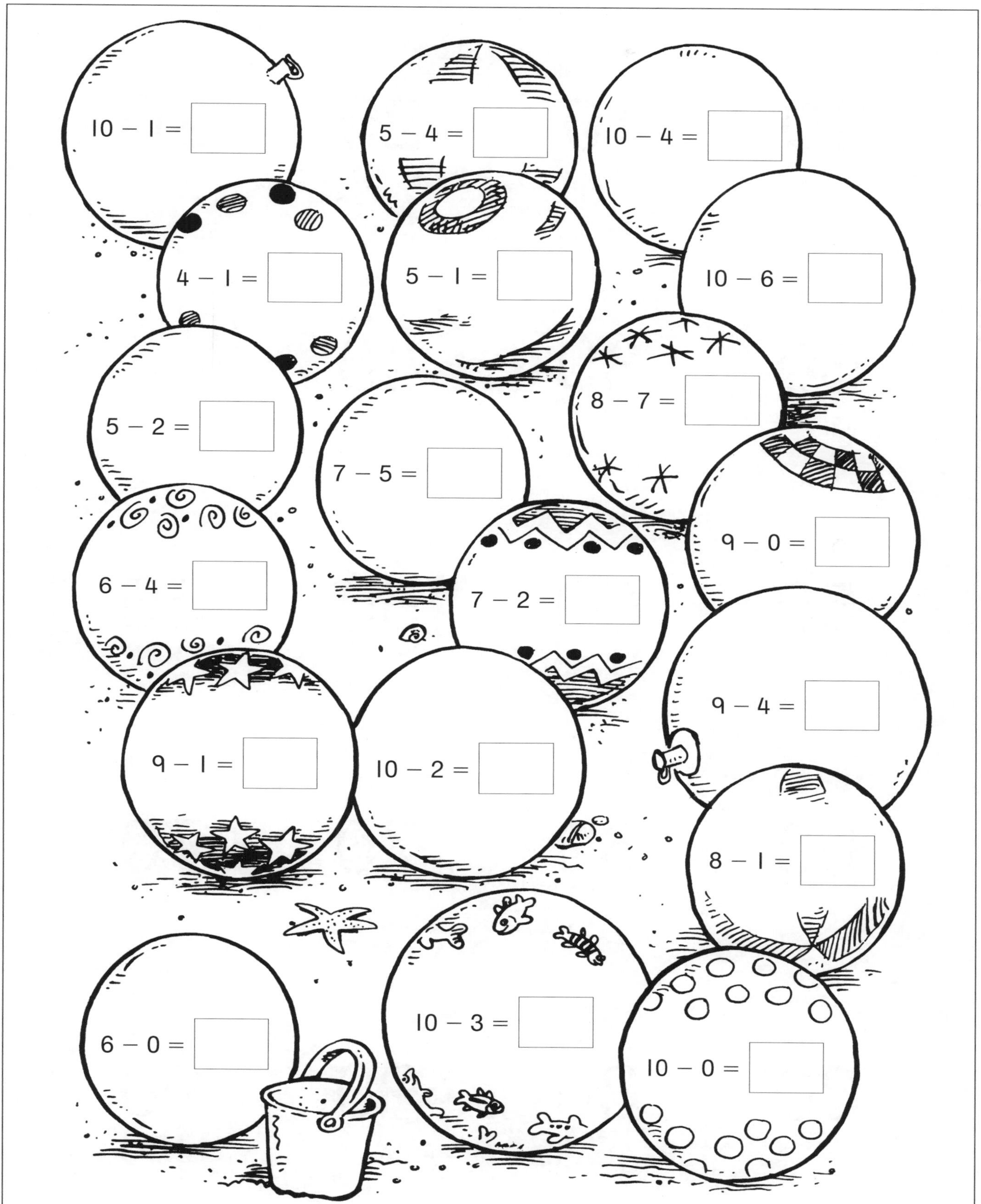

Multiples of 10 number cards

10	10	20	20
30	30	40	40
50	50	50	50
60	60	70	70
80	80	90	90

Doubling game

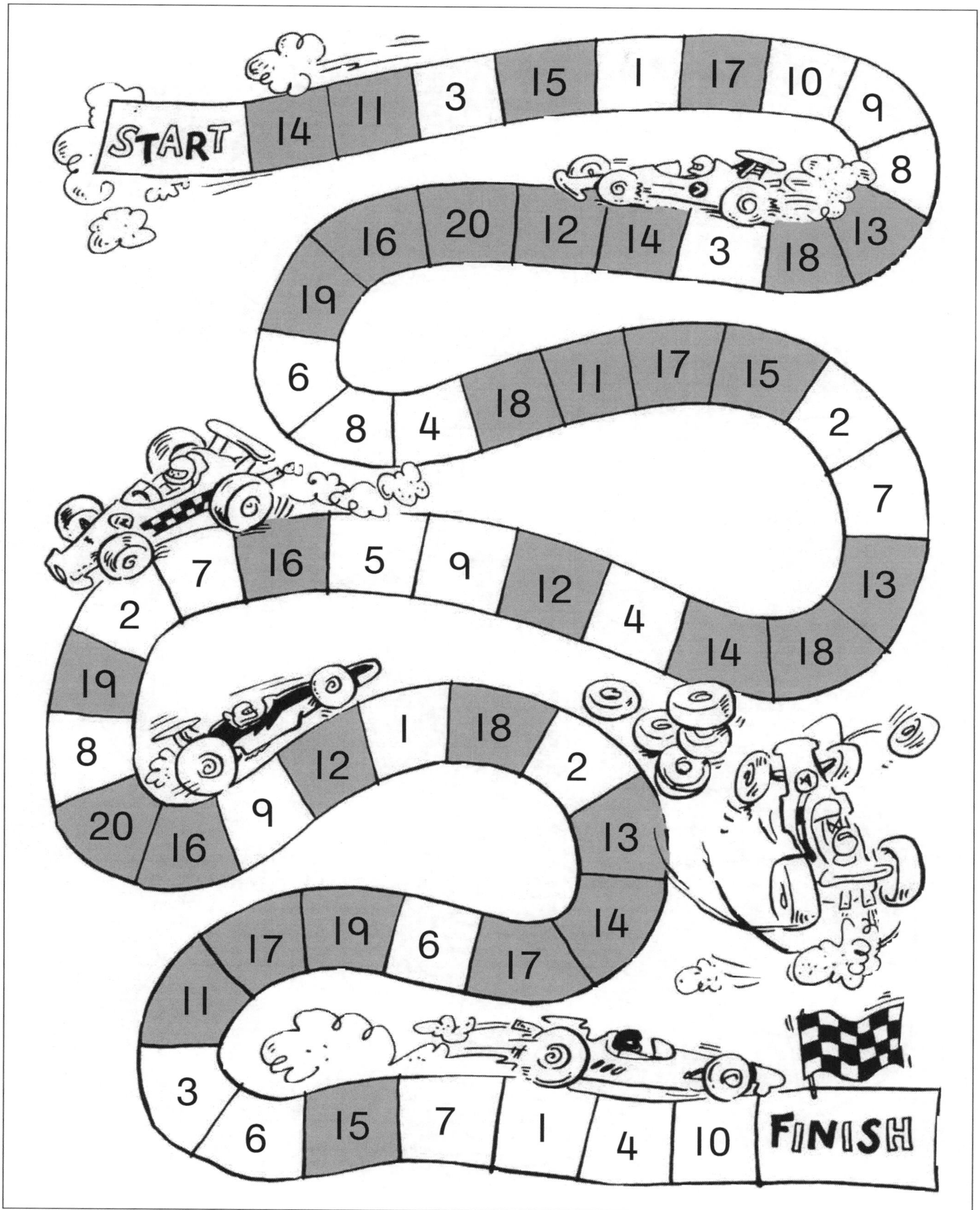

Halving game

2, 5 and 10 multiplication facts

2 times table	
14	12
6	20
18	4
2	10
8	16

5 times table	
35	40
15	25
20	30
50	45
5	10

10 times table	
90	100
70	60
10	30
50	80
20	40

3, 4 and 6 multiplication facts

3 times table	
24	18
6	27
15	9
30	21
3	12

4 times table	
24	4
32	28
8	36
16	20
40	12

6 times table	
54	6
12	24
36	42
18	30
60	48

Multiples of 2, 5 and 10

Estimate, calculate and check

	Estimation	Calculation	Check

Addition and subtraction facts

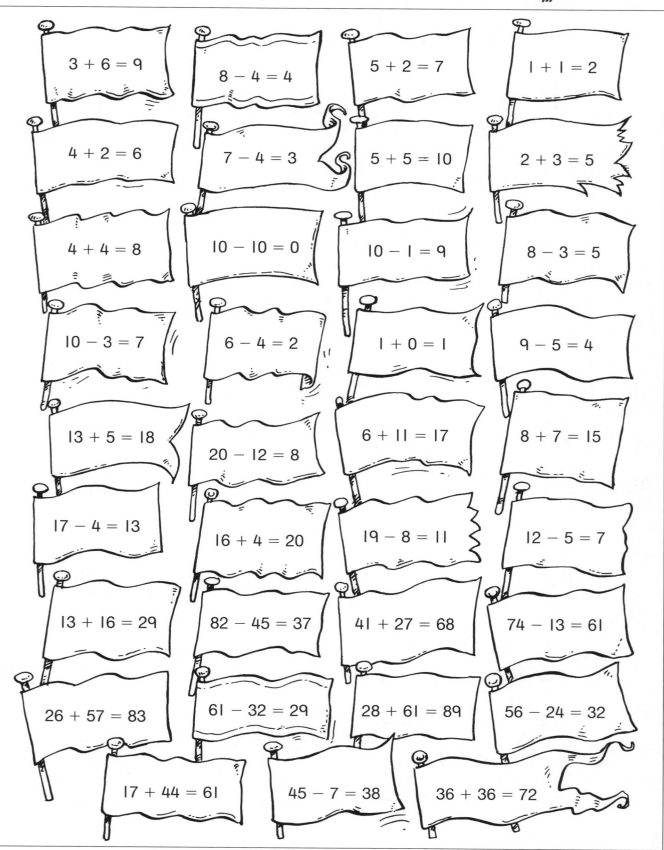

3 + 6 = 9	8 − 4 = 4	5 + 2 = 7	1 + 1 = 2
4 + 2 = 6	7 − 4 = 3	5 + 5 = 10	2 + 3 = 5
4 + 4 = 8	10 − 10 = 0	10 − 1 = 9	8 − 3 = 5
10 − 3 = 7	6 − 4 = 2	1 + 0 = 1	9 − 5 = 4
13 + 5 = 18	20 − 12 = 8	6 + 11 = 17	8 + 7 = 15
17 − 4 = 13	16 + 4 = 20	19 − 8 = 11	12 − 5 = 7
13 + 16 = 29	82 − 45 = 37	41 + 27 = 68	74 − 13 = 61
26 + 57 = 83	61 − 32 = 29	28 + 61 = 89	56 − 24 = 32
17 + 44 = 61	45 − 7 = 38	36 + 36 = 72	

Operator cards

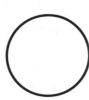

+	+	+	+
−	−	−	−
×	×	×	×
÷	÷	÷	÷
○	○	△	△

Shapes and solids

Shape property cards

Has 3 sides	Has 4 sides	Has 6 sides
Sides are straight	Sides are curved	Has 5 corners
Has 8 corners	Has 3 faces	Has 4 faces
Has 5 faces	Has 6 faces	Has a square face
Has a triangular face	Faces are flat	Has a curved face

Collins New Primary Maths

Shape cards

A

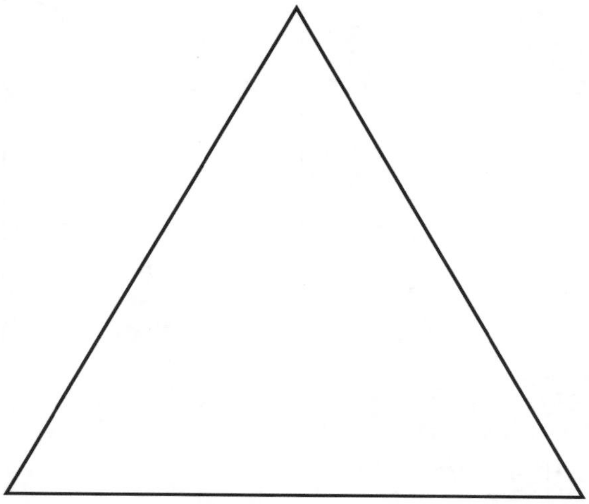

B

C

D

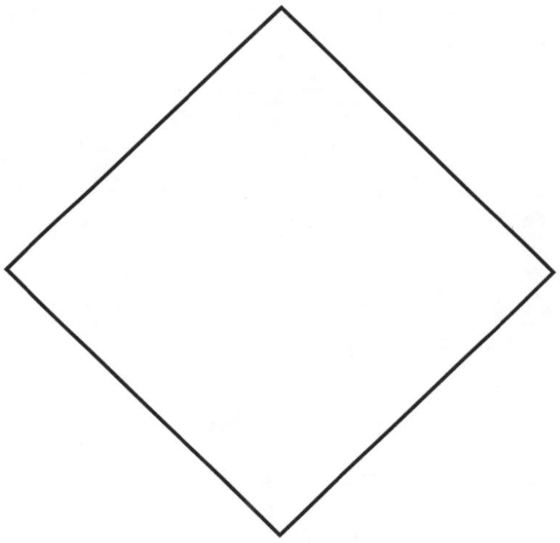

E

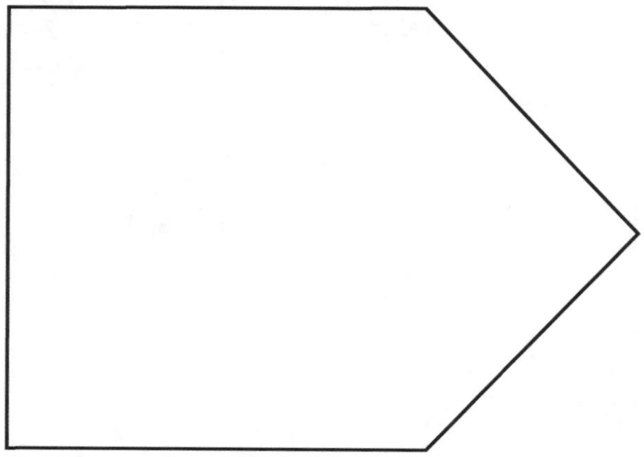

F

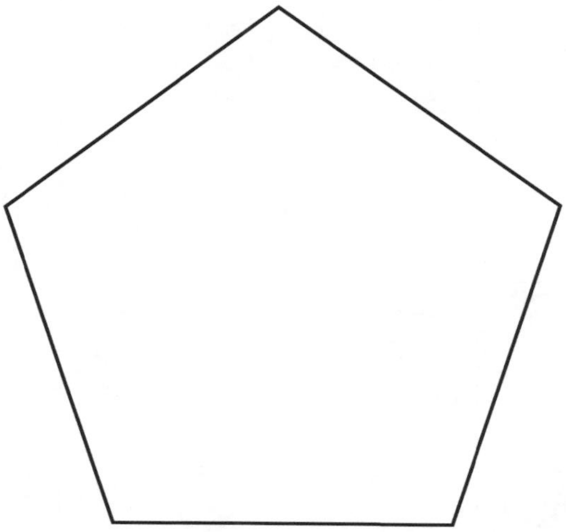

Position, direction and movement

Scales

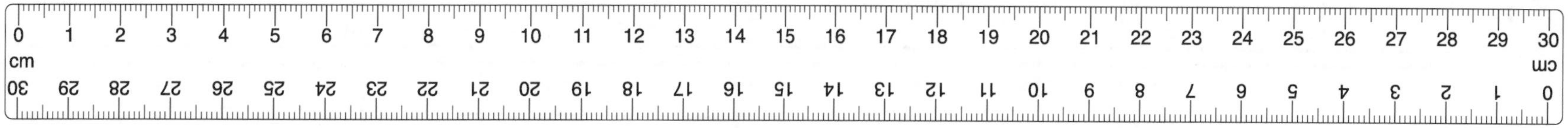
0 1 2 3 4 5 6 7 8 9 10 11 12 13 14 15 16 17 18 19 20 21 22 23 24 25 26 27 28 29 30
cm

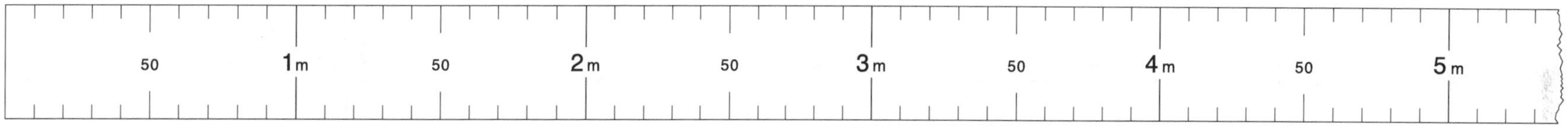
50 1m 50 2m 50 3m 50 4m 50 5m

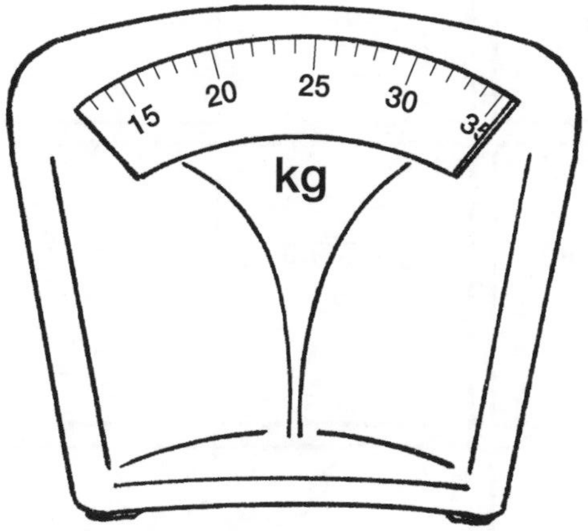
15 20 25 30
kg

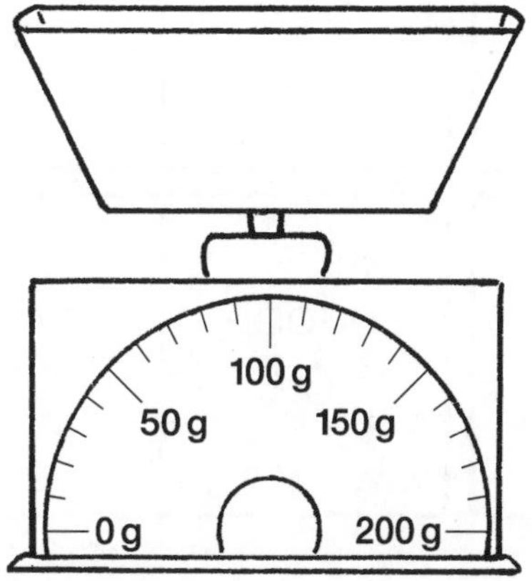
0g 50g 100g 150g 200g

litres 5 4 3 2 1

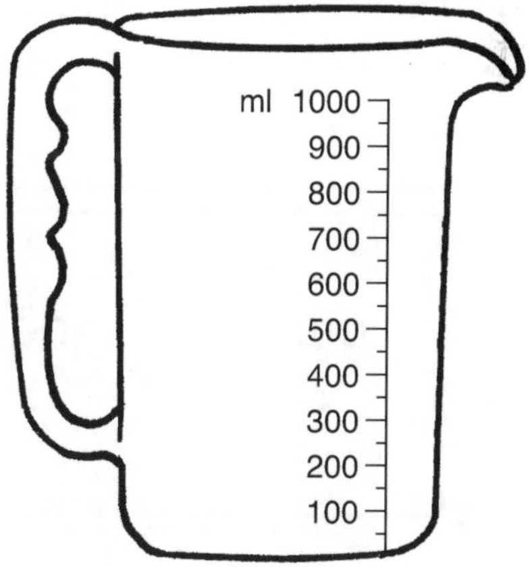
ml 1000 900 800 700 600 500 400 300 200 100

Time dominoes 1

1 week	12 months
$\frac{1}{2}$ an hour	1 year
about 30 days	60 minutes
2 months	24 hours
$\frac{1}{2}$ a year	2 days
1 minute	7 days
about 4 weeks	30 minutes
14 days	1 month
1 year	about 60 days
24 months	6 months
366 days	60 seconds
1 fortnight	1 month
$\frac{1}{4}$ of an hour	2 weeks
12 hours	52 weeks
14 days	2 years
weekend	1 leap year
1 year	2 weeks
365 days	15 minutes
1 hour	$\frac{1}{2}$ a day
1 day	1 fortnight

Time dominoes 2

48 hours	10 years	1 decade	3 minutes	180 seconds	$\frac{1}{4}$ of a year	3 months	45 minutes
$\frac{3}{4}$ of an hour	90 minutes	$1\frac{1}{2}$ hours	36 hours	$1\frac{1}{2}$ days	4 months	$\frac{1}{3}$ of a year	120 seconds
2 minutes	$\frac{3}{4}$ of a year	9 months	1 century	100 years	20 minutes	$\frac{1}{3}$ of an hour	30 seconds
$\frac{1}{2}$ a minute	$\frac{1}{4}$ of a day	6 hours	about 8 weeks	2 months	about 90 days	3 months	2 days

Collecting, recording and presenting data

Question

What information do you need to collect?

How are you going to collect the information?

How are you going to record the information?

How are you going to present the information? Why?

Now do it!

What have you found out?

If you had to answer this question again, what would you do differently? What things would you keep the same?
